Atomic Energy & The Safety Controversy

Atomic Energy & The Safety Controversy

Edited by Grace M. Ferrara

Writers: Ira Freedman, Chris Larson,
Gerald Satterwhite, Barry Youngerman

Facts On File
119 West 57th Street, New York, N.Y. 10019

Atomic Energy & The Safety Controversy

© Copyright, 1978, by Facts On File, Inc.

Library of Congress Cataloging in Publication Data
Main entry under title:
Atomic energy & the safety controversy.
  (A Facts on File publication)
  Includes index.
1. Atomic energy industries—Safety measures.
2. Atomic energy industries—Accidents. I. Ferrara,
Grace M. II. Facts on File, Inc., New York.
HD9698.A2A758   614.8′39   78-26210
ISBN 0-87196-297-7

9 8 7 6 5 4 3 2 1
PRINTED IN THE UNITED STATES OF AMERICA

Contents

Overview

FOR AT LEAST A CENTURY, THE industrialized nations have depended on fossil fuels for their energy needs. As these nonrenewable energy sources dwindle and oil prices skyrocket, however, many nations have begun to seek energy alternatives. For many, the alternative has been nuclear power.

In the United States, the possible answers include coal as well as uranium, the fuel for nuclear plants. The U.S. is fortunate in having deposits of coal and uranium considered sufficient to meet its increasing demand for electricity in this century. U.S. coal supplies are vast, but the future use of coal depends on solving pertinent economic, health and environmental problems. For that reason, the Carter Administration has decided that "the United States must count on nuclear power to meet a share of its energy deficit" (from an April 1977 White House statement on the National Energy Plan). Yet, while nuclear power seems to be the future trend in energy, it too is not without major problems to public safety, the environment and future generations.

Since the 1950s there has been public debate over the controversial issues and risks associated with the use of nuclear technology. Much of the debate is focused on the question of possible failure in the reactor core cooling systems, which could cause a melt-down in which radioactive fuel would penetrate the concrete floor into the earth below and release radiation into the environment. Public concern has also been expressed on the problems surrounding nuclear wastes (the shipment and disposal of used radioactive material), the environmental impacts of thermal pollution from the reactor cooling systems and the threat of worldwide proliferation of atomic weapons.

In response to the many questions raised over nuclear dangers, numerous studies have been made over the years, and the results derived have often been contradictory.

An important study resulted in the Rasmussen Report, sponsored by the Atomic Energy Commission (AEC). A Rasmussen draft was issued in 1974, and the final version was released in October 1975. The final version contained sections of the draft report that were completely rewritten and a section added as a result of critical comments. The study was directed by Prof. Norman C. Rasmussen of the Massachusetts Institute of Technology. This study only "considered large power reactors of the pressurized water and boiling water [all water cooled] type [used at that time.] ... Although high temperature gas cooled and liquid metal fast breeder reactor designs...[were then] under development, reactors of this type...[were] not expected to have any significant role in United States electric power production in this decade; thus they were not considered."

To date, according to industry and government sources, the safety record of nuclear reactors is excellent. Spokesmen for the nuclear power industry have asserted that no injuries or deaths have resulted from the operation of licensed nuclear plants in the U.S.

The Rasmussen Report, however, said that, "[u]sing methods developed in recent years for estimating the likelihood of...[core melt-down] accidents, a probability of occurrence was determined for each core melt accident identified. These probabilities were combined to obtain the total probability of melting the core. The value obtained was about one in 20,000 per reactor per year. With 100 reactors operating, as [was] anticipated for the U.S. by about 1980, this means that the chance for one such accident is one in 200 per year."

Frank G. Dawson asserted in *Nuclear Power: Development and Management of a Technology* (1976) that in the "first thirty-two years of nuclear power (1943-75), there [had]...been only seven deaths resulting from radiation exposure accidents. All but one of the fatal accidents were associated with experimental programs, and none was associated with commercial power reactors." Dawson also asserted that in 1966 there had been a partial core melt-down in the privately owned Fermi plant, a liquid metal fast breeder. The partial melt-down had been caused by the blockage of the sodium coolant. Liquid sodium—a highly reactive chemical, which explodes into flames if exposed to water and corrodes most metals when exposed to air—is used to cool the reactor core in this type of reactor.

Construction of the Fermi plant had begun in 1956 at the Lagoona Beach site, located just 30 miles from Detroit. By 1974, however, the plant had been decommissioned, and, according to the Nuclear Regulartory Commission, "an oil-fired boiler is now used with the plant's turbine generator to produce electricity during peak loads."

According to Dawson, the controversy over the Fermi plant began in August 1956, when the AEC issued a construction permit for the plant after ignoring its Advisory Committee on Reactor Safeguards' report "that it was not sure that the Fermi plant could be operated safely at the Lagoona Beach site." In a report by the University of Michigan Engineering Research Institute (1957), it was estimated "that a release of all radioactive material from Fermi [a 100-megawatt reactor] could result in as many as 133,000 deaths, 181,000 immediate injuries from radiation, and 245,000 long-term injuries."

McKinley C. Olson, in *Unacceptable Risk: The Nuclear Power Controversy* (1976), quoted two reports made by the Brookhaven National Laboratory, N.Y. and sponsored by the AEC. The first report, issued in 1957, estimated that many fewer casualties—3,400 deaths and 43,000 injuries—would occur from a 200-megawatt nuclear plant accident 30 miles from a city. The second Brookhaven report (1965) estimated that in an accident involving larger reactors, 45,000 people would be killed and 100,000 injured.

A decade later, the Rasmussen Report assessed the consequences of a hypothetical accident. According to *Nuclear Power Issues & Choices, Report of the Nuclear Energy Policy Study Group* (1977), a report based on a study sponsored by the Ford Foundation and directed by Spurgeon M. Keeny Jr., "an extremely serious accident under very adverse conditions is estimated. . .to kill as many as three or four thousand people over a few weeks, cause tens of thousands of cancer deaths over thirty years, and cause a comparable number of genetic defects in the next generation."

The Rasmussen Report seemed to have exacerbated the reactor safety argument rather than ended it. On June 11, 1976 the Subcommittee on Energy & the Environment of the House Interior & Insular Affairs Committee held an oversight hearing on the Rasmussen Report. One of the witnesses, Frank Von Hippel of the Center for Environmental Studies at Princeton University, offered this assessment:

...One would have hoped that, once it was out, such a study would have had a tempering effect on the debate, that the reactor safety issue would have been put into better perspective, that the area of agreement between the opponents and advocates would have been expanded, and that all parties might be talking more often in the same language instead of continuing to talk past each other.

Unfortunately, none of these things has come to pass. Instead of dampening the fires of controversy, the publication of the Rasmussen report has had the effect of adding fuel to it. Why has this happened?

In my view, the problem stems in large part from the manner in which the report was packaged. Although it contains a lot of material which can be used to put the risk from reactor accidents in perspective, it has been presented and publicized as having done something which it did not and could not do; namely, prove that the hazards from reactor accidents are negligible....

In the executive summary, the Rasmussen group's estimates of the likelihood of reactor accidents of different levels of seriousness are shown on graphs....[One graph] shows, for example, the Rasmussen group expects thousands of dam failures, each killing 1,000 people, before one reactor accident occurs with the same number of fatalities. Similarly, in a comparison of reactor accident hazards to natural catastrophes in the next figure..., the Rasmussen group concludes that the chances of 100 or 1,000 people being killed by a reactor accident are about the same as the same number of people being killed by a meteor.

Thus, it would appear from these comparisons that reactor accident hazards represent only a trivial addition to the hazards which we are exposed to every day and that those of us who worry about reactor safety should really stop wasting our time and concern ourselves with more important things.

Unfortunately, these comparisons are deceptive. They show neither the most important consequences of a reactor accident nor the great uncertainties in the calculated probabilities of their occurrence....

Much of the safety debate revolves around the fact that nuclear waste products from the past continue to pose major problems for the future. As asserted in the Ford Foundation study, "past experience with nuclear waste from the military programs and from an unsuccessful commercial venture has not been very encouraging." Over more than 35 years, the federal government has stored 205 million gallons of liquid high-level waste. Currently, this waste is being temporarily held at federal installations. However, between 1958 and 1974, 19 leaks totaling 429,500 gallons have been reported at these sites. In one case, in 1973, 115,000 gallons leaked out undetected for 48 days at the Hanford site in Richland, Wash.

Commenting on this problem, Sen. Charles M. Mathias Jr. (R, Md.) said in a Senate speech Oct. 10, 1977: "The people of America have learned that the benefits and promise of nuclear power... involve both expense and risk....Congress now has the obligation to reduce the risk by addressing one of the most serious hazards of nuclear development—the containment and management of nuclear waste, both military and commercial....The challenge is

great, because this deadly radioactive residue must be totally isolated from the environment for hundreds of centuries. If left un-
checked...the menace of nuclear waste...will grow and pose a
dangerous threat to the very survival of mankind....Congress has
appropriated hundreds of millions of dollars for research and development of waste management techniques...for many years...
but the incredible fact is that the technical solution for long-term
disposal of nuclear wastes recently recommended to Congress—
sealed storage in stable geologic formations such as salt-beds—is
the same one made to the Atomic Energy Commission by the National Academy of Science 20 years ago.''

Government leaders, environmentalists and other concerned
citizens continue to debate the safety-related issues that are discussed in this book. Through public intervention, court battles and
government reform, these issues have become recognized as a national and world problem. This book records some of the more important events and actions of the 1970s associated with the nuclear
safety controversy. The material in this book is taken largely from
the printed record compiled by FACTS ON FILE in its weekly
coverage of world events. A conscientious effort was made to record all developments without bias.

GRACE M. FERRARA

October, 1978

Safety Controversy

Safety Systems in Question

The decades-long dispute over nuclear hazards to human health and the environment continued on into the late 1970s. Nuclear-power projects in existence and on the drawing board were denounced by scientific and environmental groups as well as some members of Congress as unduly threatening the future of mankind. Supporters of nuclear power insisted that atomic energy was safe as well as necessary for the continuation of industrial progress.

Safety equipment criticized. Concern within the Atomic Energy Commission (AEC) that standard nuclear power plant safety equipment might be inadequate surfaced during hearings begun by the agency Feb. 2, 1972, in response to criticism by environmental and scientist groups.

Philip L. Rittenhouse of the AEC's Oak Ridge (Tenn.) National Laboratory testified March 10 that he and 28 other nuclear safety experts, many of them AEC employes, questioned the efficacy of the emergency core cooling system. The system was designed to flood nuclear reactors with cool water if mechanical failure caused reactor temperatures to rise dangerously. A cooling failure could theoretically cause an atomic explosion.

As a safety measure, the Commission in 1971 had ordered five existing power plants to modernize their emergency cooling systems and three plants to lower their peak operating temperatures to 2,300 degrees Fahrenheit. The orders were described as part of an "interim core cooling policy," adopted after laboratory models of the cooling system had failed to function properly despite computer predictions of success.

A-safety probe asked. A coalition of 60 environmental groups, the Consolidated National Intervenors, asked the U.S. Congressional Joint Atomic Energy Committee Sept. 13, 1972 to begin a "full-fledged, in-depth" study of safety problems in atomic energy plants that had been debated in eight months of still uncompleted hearings before the Atomic Energy Commission.

The problems centered on the reliability of the emergency cooling system. The Intervenors criticized the AEC for issuing two new operating licenses while the controversy continued, although the commission had ordered reductions in maximum permissible operating temperatures for the 25 previously licensed plants. A spokesman for the environment group claimed that a West German panel of scientists equivalent to the AEC Safety and Licensing Board had asked for a moratorium on new licenses in that country.

The Intervenors had also asked the AEC Feb. 1 to suspend all nuclear power plant licensing pending full-scale tests of the cooling system, scheduled for 1975.

Sen. Mike Gravel (D, Alaska) had introduced a bill Feb. 23 to halt all operations and construction of nuclear plants until the "probabilities of major accidents and nuclear pollution are reduced by tested methods, until the justification for creating a permanent radioactive legacy is more widely debated and until alternative energy sources are considered."

Dr. Edward E. David Jr., presidential science adviser, told the Congressional Committee Sept. 12 that atomic plants, and in particular fast breeder reactors, were the most promising source of pollution-free energy for the next 20 years. He said in addition that research in nuclear fusion as an energy source had progressed far enough so that the program would be expanded beyond this year's $61 million level, but he reported no important gains in coal gassification research.

The AEC had reported to the committee Sept. 7 that it had committed itself to pay all costs of developing the fast breeder reactor above the $250 million it expected private industry to contribute. Costs, including development and five years of operation, were expected to be about $700 million.

Breeder reactor opposed. A group of 31 scientists and other experts in a statement April 25, 1972, asked Congress to deny an Administration request for development funds for an experimental fast breeder reactor power plant because of safety and environmental hazards.

The group, including Dr. Linus Pauling, Dr. Harold C. Urey, Barry Commoner, Dr. Paul Ehrlich and Dr. George Weil, warned that the heat generated in a breeder reactor could lead, in cases of human or mechanical error, to a substantial atomic explosion. Problems of plutonium transport and disposal were also raised, as well as the danger of black market plutonium diversion for weapons manufacture.

Weil, the group's spokesman, urged that coal be reconsidered as an alternative energy source, since "it can be cleaned up, hopefully."

The AEC had defended the breeder reactor April 14 in an environmental impact statement for the proposed demonstration plant, to be built at an eventual cost of $2 billion–$3 billion as part of the Tennessee Valley Authority power grid. (The AEC had refused to prepare an impact statement for an entire generation of plants, as demanded by environmental groups.) The agency claimed that coal mining disrupted as much as 4,000 times as much land as equivalent nuclear power systems. The breeder reactor would produce additional fuels as it operated.

Doubts about the reactor were expressed by the Interior Department, which said April 14 that the danger of a "hydrogen explosion" had been insufficiently explored.

A-plant moratorium asked. At a Jan. 3 1973 Washington news conference, Ralph Nader and spokesmen for the Union of Concerned Scientists asked the Atomic Energy Commission (AEC) for a moratorium on all construction of new atomic power plants "until all safety-related issues are resolved." Nader promised to continue his compaign to delay massive use of atomic power, on economic and safety grounds in Congress, the courts and among electric company stockholders.

Nader and the scientists' group said the danger of "catastrophic nuclear power plant accidents is a public safety problem of the utmost urgency," and called for output reductions of up to 50% at the 29 operating nuclear reactor power plants. Although "no nuclear explosion could occur," they said, a failure in the emergency core cooling system could cause release of radioactive materials with deaths possible nearly 100 miles from the plant. Nader also said that disposal of radioactive wastes presented safety problems.

Nader charged the AEC with secrecy on the issue and with failure to publicize what he said was a belief held by a majority of AEC scientists that safety problems may exist.

In reply, William R. Gould, chairman of the Atomic Industrial Forum, said in New York that "throughout the civilized world" there was a "massive" shift to nuclear energy as a power source "because of its advantages in terms of fuel supply, economics, environmental affects and public health and safety." He said he was confident that "the extraordinary safety record of nearly 100 operating power reactors worldwide" and "the extremely conservative engineering approach" of U.S. plants would be confirmed at February hearings of the Congressional Joint Atomic Energy Committee.

A-plant suit rejected. Federal Court Judge John H. Pratt June 28, 1973 rejected a suit to close 20 atomic units because of alleged safety hazards. The suit had been filed May 31 by consumer advocate Nader and Friends of the Earth, an environmentalist group. The suit had charged the AEC with a "gross breach" of its public health and safety obligations by failing to assure that power plant cooling systems would not fail and cause the release of radioactive materials.

David Brower, of Friends of the Earth, had said that "overwhelming scientific evidence" had shown that "the lives of millions of people" were threatened by operation of the plants because of "crude and untested" safety systems. The suit quoted several AEC officials as expressing doubt as to the reliability of back-up cooling systems.

In a statement on the suit, the AEC had said it saw no basis for suspending operation of the units, which were at 16 power plants in 12 states. The AEC conceded there were differences in opinion within the commission on the safety systems but noted that a review was already under way to determine whether present core-cooling regulations were adequate.

Judge Pratt ruled that the plaintiffs had not exhausted AEC procedures for challenging nuclear power plant safety. Pratt also said the AEC was making a proper investigation.

A renewed petition was rejected by the AEC Aug. 30. In doing so, the AEC stood by its earlier position that existing interim regulations covering core-cooling systems provided "reasonable assurance" of protection of public health and safety.

Tighter safety rules imposed. The AEC voted Dec. 28, 1973 to impose stricter safety standards on heating of nuclear fuel and operation of emergency cooling systems in nuclear power plants.

The new rules set the maximum temperature for operation of uranium fuel piles in the 26 existing water cooled plants at 2,200 degrees Farenheit, a reduction of 100 degrees from the old standards.

The rules would require power output reductions of up to 20% at some plants, and could lead to extensive modifications in plants now operating or under construction. The Washington Post cited an industry source Oct. 27 as estimating that the rules would cost the industry $50–$100 million over the next five years.

Another rule required that fuel bundles be redesigned to insure that no more than 17% of the reactor-core shielding would oxidize if it came in contact with cooling water.

The commission gave the industry six months to comply with standards.

The Union of Concerned Scientists, an intervenor in the AEC proceeding, said Dec. 28 that the ruling was "cosmetic" and represented a "continuation of the AEC's cover-up of critical safety problems."

The rules, which were expected to increase energy costs, had been proposed Oct. 26, 1972 by L. Manning Muntzing, the AEC's director of regulation.

The purpose of the change was to reduce the risk of accidental radiation leakage.

Power plant cutbacks ordered—The AEC Aug. 24, 1973 ordered 10 nuclear plants in seven states to cut back power levels 5%-25% pending studies of possible safety hazards.

The AEC said the precaution was taken because of the discovery of shrinkage in uranium oxide pellets in reactor fuel rods. With the shrinkage, the heat of the atomic process was not efficiently transferred to cooling water, narrowing the safety margin in case of cooling system failure. The shrinkage could also cause collapse of fuel rods.

The cutbacks would be in effect until the commission could evaluate new data from General Electric Co., manufacturer of the reactors.

The Federation of American Scientists had asked the AEC Feb. 6, 1972 to order a reduction in nuclear reactor operating levels, as the agency had proposed in 1972, to avoid an accident that "could mean death to tens of thousands of people" through radiation release.

Although such an accident was unlikely, the federation said, the AEC should undertake "a crash program of stepped up reactor safety research," especially concerning "alternative reactor systems," which could include gas-cooled reactors. Nearly all current reactors, and most planned reactors, were water-cooled,

New Agencies Replace AEC

AEC replaced, ERDA established. A bill establishing the Energy Research & Development Administration (ERDA) was approved by the House Oct. 9, 1974, and the Senate Oct. 10 and was signed by president Gerald R. Ford Oct. 11. The ERDA, which would consolidate federal energy research, replaced the Atomic Energy Commission (AEC), which was abolished under the legislation. The AEC, with a $1.7 billion fiscal 1975 budget, oversaw a $10 billion complex of national laboratories.

The AEC's licensing and regulation activities would be handled by a new Nu-

clear Regulatory Commission. The bill specified that the heads of the three regulatory offices—reactors, materials safeguard and research—would have direct access to the commissioners of the new agency. The commissioners were required to make public disclosure of any safety-related "abnormal occurrences" at nuclear plants within 15 days. Manufacturers and nuclear facility officials were required to report equipment or operating defects threatening public safety. A maximum penalty of $5,000 was set for each individual violation.

The nuclear weapons development aspect of the AEC operation was tentatively put under the aegis of the new agency, which, along with the Defense Department, would prepare recommendations for the future disposition of the responsibility.

An Energy Resources Council (formerly the National Energy Board) authorized under the legislation was established by President Ford by executive order Oct. 11, with Interior Secretary Rogers C. B. Morton as chairman.

Debate over U.S. Radiation Standards

As the safety argument continued, federal radiation standards became a target for debate. Government officials voiced conflicting views over whether or not the allowable radiation emissions from nuclear plants should be reduced to protect public health. One result was that early in 1977 the Environmental Protection Agency lowered the radiation limit from 500 millirems to 25 millirems per person annually, and ordered further reductions to be made by 1983.

Nuclear discharge standards debate. A government radiation expert for atomic energy said Sept. 6, 1970 that "hazards from nuclear power plants are being badly over-painted."

The opinion was voiced by Paul

Tompkins, executive director of the Federal Radiation Council, the body responsible for setting nuclear discharge standards, and was endorsed by Saul Levine, assistant director of the Atomic Energy Commission's (AEC) Office of Environmental Affairs.

(The AEC was later replaced by the ERDA.)

AEC proposes rules change. Atomic Energy Commissioner James T. Ramey said March 27 that the AEC was proposing regulations to require industry to take advantage of improvements in technology to minimize the amount of radiation released by new nuclear plants. (Current regulations required only that emissions of radiation be within AEC safety limits.)

Ramey said the proposed rules would not reduce the current maximum permissable limits of radioactivity emitted from nuclear power plants. He denied there was any connection between the proposed changes and a recent case in Minnesota in which the state sought limits on radioactive discharges that would be more rigorous than the AEC's.

Supreme Court rules on radiation—The Supreme Court ruled April 3, 1972 that the AEC alone had the right to regulate radiation control standards for nuclear power plants.

The decision dealt a serious setback to efforts by local governments to seek tighter safety controls.

The court affirmed a lower court ruling involving the state of Minnesota and a nuclear power plant built by the Northern States Power Co. 30 miles north of Minneapolis.

Northern States had successfully sought an AEC operating license that would have permitted a daily discharge of 41,400 curies of radioactive debris. A curie was a unit of radioactive disintegration.

That discharge, however, would have violated a limit of 860 curies a day set earlier by Minnesota's State Pollution Control Agency.

Two lower U.S. courts had sustained Northern States' contention that it would have been either impossible or prohibitively expensive to meet Minnesota's limits. More than a dozen states had sided with Minnesota's position during the litigation.

Chief Justice Warren Burger, and Justices Harry Blackmun, William Brennan Jr., Thurgood Marshall, Lewis Powell, William Rehnquist and Byron White made up the majority. Dissenting were Potter Stewart and William Douglas.

AEC reduces radiation limit. The AEC June 7, 1971 proposed that A-plants limit their radiation exposure to 1%, or less, of the amount of radiation permitted under current U.S. guidelines for power reactors. AEC spokesmen said only the Humboldt Bay reactor near Eureka, Calif., and the Dresden reactor no. 1 and possibly the Dresden reactor no. 2, both near Chicago, failed to meet the new guidelines.

Radiation levels reported safe. A study released Jan. 25, 1971 by the National Council on Radiation Protection (NCRP) in Washington said current precautions against radiation exposure in the U.S. were safe and no tighter standards were required. The council was a nonprofit corporation chartered by Congress and provided information on radiation projects of government agencies, including the Atomic Energy Commission and the Public Health Services. Its recommendations were to be submitted to the Environmental Protection Agency.

The council report, based on a 10-year study, said its "review of the current knowledge of biological effects of radiation exposure provides no basis for any drastic reductions in the recommended exposure levels despite the current urgings of a few critics."

The NCRP took issue with recent assertions by Drs. John F. Gofman and Arthur R. Tamplin of the AEC's Radiation Laboratory in California. According to their own estimates of radiation dangers, the two scientists concluded that 32,000 excess cancer deaths would probably occur annually under the average permissible radiation exposure (170 millirems) of the American population. Gofman urged that the radiation exposure limits be reduced to one-tenth their present levels.

The NCRP contended that exposure to the total permissible maximum would result in no more than 3,000 additional cancer deaths a year. It said that although average exposures were far lower than the recommended limit, no exposure could be regarded as safe.

The council recommended for the first time that a maximum permissible radiation dose of 0.5 rem (measurement of radiation dosage) be set specifically for pregnant women to protect the fetus.

EPA radiation authority curbed. Charging that the Environmental Protection Agency (EPA) had "construed too broadly" its responsibilities to set environmental standards on radioactive materials, the White House ordered the EPA to drop its plans to set radiation rules for individual nuclear power plants, the New York Times reported Dec. 12, 1973.

According to the directive, the authority to set such standards would rest solely with the Atomic Energy Commission (AEC), which was considered by critics of the nuclear power industry to have a more lenient approach to radiation protection than the EPA.

The order was in a memorandum, dated Dec. 7, sent "on behalf of the President" by Roy L. Ash, director of the Office of Management and Budget, to EPA Administrator Russell E. Train and AEC Chairman Dixy Lee Ray. The memo said its purpose was to prevent "confusion" in the area of nuclear power regulation, "particularly since nuclear power is expected to supply a growing share of the nation's energy requirements."

Ash said the AEC should proceed with its own plans to issue rules and the EPA should discontinue similar preparations to issue standards "now or in the future." The EPA would retain responsibility "for setting standards for the total amount of radiation in the general environment from all facilities combined in the uranium fuel cycle"—flexible standards which "would have to reflect the AEC's findings as to the practicability of emission controls."

The Times quoted Charles L. Elkins of the EPA's office of hazardous materials control as feeling that AEC standards were "not adequate."

EPA sets new limit. The Environmental Protection Agency set a new limit Jan. 6, 1977 for the amount of radiation emissions permitted from the normal operation of nuclear facilities.

The limit, in terms of exposure for any member of the general public, would be a maximum annual radiation dose of 25 millirems to the whole body (75 to the thyroid gland).

The new standard replaced an advisory "radiation guide" with a limit of 500 millirems to the whole body (1,500 to the thyroid gland). Unlike the advisory guide, the new standard was legally enforceable, according to the EPA.

The new standard also called for reduction of allowable emissions of krypton-85 to one-tenth of current levels by 1983.

EPA Administrator Russell E. Train described the new standards as "an important precedent in radiation protection because they consider the long-term potential buildup of radiation in the environment."

The EPA said most nuclear power plants already conformed to the new standards but that milling and other fuel-supply and reprocessing operations needed to be upgraded to meet the new restrictions.

The tighter standards reflected an assessment of the impact of environmental radioactivity for this and future generations.

Environmental Issue

Congress in 1970 had passed the National Environmental Policy Act (NEPA) as a legal tool to promote the safety of nuclear plants. The NEPA required all federal agencies to file an environmental impact statement for each projected nuclear plant.

AEC's statement disputed. The Environmental Protection Agency Feb. 16, 1973 rejected as "inadequate" an AEC environmental impact statement on emergency safety procedures for nuclear power plants.

EPA said AEC should have explored the possibility of catastrophic accidents, such as the loss of coolant or pressure vessel failure, and suggested an independent study be commissioned in cooperation with both agencies. The environmental agency conceded that no serious accident had occurred in a nuclear plant, but said the number of such plants in the U.S. would rise from 30 at present to 1,000 by the year 2000.

According to a Feb. 27 report, an AEC spokesman said the agency was not prepared to discuss catastrophic accidents until completion of an internal study which was expected to take another year.

A three-man AEC board held hearings Feb. 1–4 on environmental issues relating to power plants, including problems in mining, refining, transport and disposal of radioactive fuel. The hearings were designed to help speed licensing procedures for the 120 power plants already planned or begun by considering at one time the problems common to all the plants. Dr. Henry J. Kendall, a nuclear physicist and member of the Union of Concerned Scientists, said at the Feb. 1 hearing that the AEC had published estimates of a one in 1,000 chance of a major reactor pipe break, a risk he considered "totally unacceptable."

NEPA amendment proposed. The AEC, backed by the top Administration environment officials, had asked Congress March 16, 1972 for an amendment to the National Environmental Policy Act to permit the AEC to grant interim emergency operating licenses to new nuclear power plants prior to filing complete environmental impact statements. After passage in the House, the bill died in the Senate.

AEC Commissioner James R. Schlesinger told the Joint Committee on Atomic Energy that without emergency licenses, five completed nuclear plants would be unable to meet possible 1972 and 1973 power shortages in New York City, Illinois, Iowa, Michigan and Wisconsin.

Under the proposed amendment, the AEC would have the power through June 1973 to issue licenses without public hearings and without submitting statements to the Council on Environmental Quality. After July 1, 1973 no new interim licenses would be issued, but any license could be continued if the AEC declared the emergency still in effect.

Russell E. Train, chairman of the Council on Environmental Quality, and William D. Ruckelshaus, administrator of the Environmental Protection Agency, testified in favor of the amendment, leading some congressmen to charge that the Administration was planning to cripple the environmental act.

Illinois license granted—The most critical energy situation that would be covered by the amendment, Schlesinger said, concerned the Quad Cities dual-reactor plant at Cordova, Ill., whose license had been blocked by a federal judge at the request of the State of Illinois, the Izaak Walton League and the United Auto Workers.

The plaintiffs agreed March 30 to drop their objections when the companies, the Commonwealth Edison Co. and the Iowa-Illinois Gas & Electric Co. agreed to build a four-mile $30 million cooling system to prevent thermal pollution of the Mississippi River. Testing licenses for the plant were issued March 31.

Schlesinger said March 30 that the accord "alleviates the immediate pressure" for new legislation, but he still hoped for Congressional action to facilitate licensing of at least 13 other nuclear plants under construction.

AEC breeder impact report ordered. A federal appeals court in Washington ruled June 12, 1973 that the AEC must prepare a formal environmental impact statement for the entire program of projected liquid metal-fast breeder reactor nuclear power plants, envisioned as the eventual answer to growing power needs.

The ruling overturned a lower court decision by Judge George L. Hart, Jr., who March 24, 1972 had dismissed a suit brought by the Scientists' Institute for Public Information to force the AEC to assess the long-term effects of breeder reactor use as required by the National Environmental Policy Act of 1970 (NEPA). The lower court had ruled against the institute because of a lack of tangible impact in the near future.

The institute attorney, J. Gus Speth, had said that the type of reactor planned by the government would entail greater risk of explosion, thermal pollution and plutonium emission, and produce more radioactive wastes, than other types, and argued that all the problems should be discussed in public before the program's momentum and budget commitments made it inevitable.

Speth said annual federal expenditures of $100 million and the AEC plan to give the breeder reactor top priority together constituted a major federal act and a legislative proposal, thereby invoking the impact statement requirement.

Writing for the appeals panel, Judge J. Skelly Wright said the program presented "unique and unprecedented environmental hazards." The institute had argued that since such reactors "breed" plutonium—an extremely toxic radiological metal—and that construction of over 1,000 such plants had been projected, the potential harm of such systems should be studied before construction started.

Although the AEC had issued an impact statement for a demonstration reactor near Oak Ridge, Tenn., Judge Wright said the commission had taken an "unnecessarily crabbed approach to NEPA in assuming that the impact statement was designed only for particular facilities rather than for analysis of broad agency programs."

Wright noted that the proposed plants were expected to generate some 600,000 cubic feet of high-level radioactive wastes by the year 2000 and said the problems

"attendant upon processing, transporting and storing these wastes, and the other environmental issues raised by the widespread deployment [of such plants] warrant the most searching scrutiny under NEPA."

The decision also required that, in addition to the environmental impact report, the AEC make "a detailed statement" on alternatives to the breeder reactor program.

Reactor Safety Studies

In an effort to answer the critics of nuclear power, the AEC initiated a reactor safety study to estimate the accident risks associated with nuclear plants. The study was directed by Dr. Norman Č. Rasmussen of the Massachusetts Institute of Technology. A draft of the Rasmussen report was published in August 1974 and the final report was released in October 1975 under the auspices of the Nuclear Regulatory Commission.

Rasmussen draft disputed. According to the Rasmussen draft report on reactor safety, released by the AEC Aug. 20, 1974 the risks from nuclear power plant accidents were considered to be much smaller than from other man-made or natural disasters. The study concluded that, given the 100 conventional water-cooled plants expected to be in operation by 1980 (51 were currently operating), the chance of an accident involving 10 or more fatalities was one in 2,500 a year; an accident involving 1,000 or more deaths carried a risk of one in one million a year.

AEC officials emphasized that the study dealt only with the safety of the commercial reactors themselves, and not with risks involved in mining, manufacturing or transporting nuclear materials. The draft report also excluded the liquid-metal-fast breeder reactors currently under development.

The report stated that of the approximately 15 million persons expected to be living in the vicinity of the first 100 reactors, one might be killed and two injured in every 25 years. The study noted that power plant reactors could not explode like nuclear weapons because of the fuel used.

AEC Chairman Dixy Lee Ray said Aug. 20 that there was "no such thing as zero risk," but in terms of the study the nuclear industry "comes off very well." She said the risks cited in the study were acceptable and urged that plant construction be continued. Ray said the study had been commissioned by the AEC, but the agency did not "influence" its findings.

Reactions to AEC study—Reactions by nuclear power critics were less favorable. Daniel Ford of the Union of Concerned Scientists contended that the draft report was, "for all practical purposes," an "in-house" effort by the AEC, which had consistently advocated increased nuclear production. Consumer advocate Ralph Nader labeled the report "fiction," expressing doubt that such statistical projections were valid or possible.

Both Nader and Ford asked why—if the industry was so safe—it did not accept full liability for payment of damages in the event of accidents.

The Rasmussen draft report was also criticized in a document released Nov. 23 by the Union of Concerned Scientists and the Sierra Club, a conservationist group, as speculative and unreliable.

EPA doubts report. The Environmental Protection Agency Dec. 4 lauded the methods used by the Rasmussen study on the possibility of a serious reactor accident but said the projected casualties could be "about 10 times higher than those estimated" in the AEC survey.

The EPA statement was contained in an AEC-requested analysis of the commission's 14-volume study completed in the summer. It said the commission's study was "an innovative forward step in risk assessment of nuclear power reactors."

A second report issued by the EPA expressed concern about the possible effects the plutonium-fueled reactors proposed by the AEC would have on the environment. Before this new fuel was used on a full scale, the problem posed by plutonium disposal should be resolved and a new accident survey should be completed, the EPA said.

Speaking for the two groups, Dr. Henry Kendall, a physicist, said in Washington that the AEC's safety claims "are a conceit based far more on their enthusiasm for the nuclear power program than on solid and convincing scientific proof." The UCS and Sierra statement said the AEC study prepared by Dr. Norman C. Rasmussen contained a number of flaws. The safety analysis used by Rasmussen to estimate the probability of an accident, they said, had been developed and then abandoned by the aerospace industry and the federal government because it had been found to drastically underestimate existing hazards.

The AEC report was also criticized for the low number of projected casualties based in part on the successful evacuation of persons living near an atomic plant. A major accident at a nuclear plant could kill or seriously injure 126,800 people, 16 times the casualties estimated by the AEC report, the scientists said. They emphasized that there were no "adequate plans or means to evacuate to a distance of 20 miles" in the event of a plant mishap. It could not be assumed, as the Rasmussen report contended, that an evacuation could be achieved while only 5% of the population was in automobiles and 90% were indoors.

The report said the AEC study also failed to consider the fact that plutonium, the more lethal fuel which industry hoped soon to use in its reactors, posed a far greater danger than that considered in the commission's own reactor safety study.

AEC member Saul Levine defended the Rasmussen report Nov. 24, saying that the scientists' criticism of the study's form of analysis would have been correct for the 1960s, but "we have advanced that methodology considerably ... our members are in touch with reality." The AEC, he said, "is confident that the techniques used in the Rasmussen report are the best available."

Physicists study reactor safety. A study made by the American Physical Society April 28, 1975 found that there was no reason for "substantial short-term concern" about the safety of U.S. nuclear reactors. But it recommended "a continuing effort to improve reactor safety as well as to understand and mitigate the consequences of possible accidents."

The study, by 12 independent physicists, found the safety record so far "excellent," with "no major release of radioactivity." But the group did express concern about the long-term, when an increasing number of reactors would be operating and the likelihood of an accident, although "improbable," became correspondingly greater.

The physicists agreed with the AEC estimate that the chance of a major accident in an atomic power plant was one in 10 million. But they disagreed with the AEC's estimate that an accident could cause about 310 deaths from cancer. There would be "substantial long-term consequences" from an accident in which radiation was released over a populated area, the panel thought. The AEC estimate was said to be off by as much as a factor of 50 because the cloud of radiation could move 500 miles and endanger people in a 10,000–20,000 square-mile area.

Final Rasmussen report. After considering the comments and recommendations evoked by the draft Rasmussen report, the Nuclear Regulatory Commission (NRC) in October 1975 issued a revised report called the Reactor Safety Study.

In its general conclusions, the final Rasmussen report asserted that "[t]he results of the Reactor Safety Study indicate that nuclear power plants have achieved a relatively low level of risk compared to many other activities in which our society engages. Although the study has developed some insights that contribute to a better understanding of reactor safety, the existing low level of risk has been achieved principally by the efforts of industrial design, construction and operation and by the efforts of the AEC's regulatory process...."

"Decision making processes in many fields, and especially in safety, are quite complex and should not lightly be changed. This is especially true where a good safety record has already been obtained, as is so far true for nuclear power plants."

The report conceded that "[o]ne of the first questions that arises about the results of the study concerns its applicability to reactors other than those studied. There are those who will question the extension of results beyond the two reactors involved in the study; there are also those who will try to extrapolate the results to 1,000 reactors. The reactors studied are the 24th and 34th large reactors to come into operation. Their designs were started in 1966. The 100th plant is expected to come into operation in about 1981; its design started in about 1971. The 1,000th plant is not yet a concept; nor is it clear that 1,000 water reactors of the types studied will be built.

"By the same token, the first 100 plants, although they involve some detailed differences in design, all meet similar safety requirements and generally have the same types of engineered safety features. Thus, the extrapolation of these results to 100 reactors seems fairly reasonable. It will also tend to overestimate rather than underestimate the risks involved, because significant improvements were made in AEC's safety design requirements, in the implementation of these requirements, and in the applicable codes and standards used in the design of nuclear power plants in the years between 1966 and 1971.

"The study devoted a significant effort to ensuring that it covered the potential accidents important to the determination of public risk. In its analysis of potential nuclear power plant accidents, the Reactor Safety Study relied heavily on the twenty or more years of experience that exists in the analysis of reactor accidents. It also went considerably beyond the conventional analyses performed in connection with the licensing of reactors by considering failures that are not normally covered in standard safety evaluations.

Thus, in addition to defining the various initiating events that might potentially cause accidents, the study estimated the likelihood and consequences of the failure of the various engineered safety features provided to prevent accidents and to cope with the consequences of accidents. Failures of reactor vessels and steam generator vessels as potential accident initiators were considered. The availability of systems to remove decay heat from a shutdown reactor was examined as an additional part of the assessment of transient events. The likelihood that various external forces might cause reactor accidents was also taken into account.

"The following factors provide a high degree of confidence that the significant accidents have been included: 1) the identification of all significant sources of radioactivity located at nuclear power plants, 2) the fact that a large release of radioactivity can occur only if reactor fuel melts, 3) knowledge of the factors that affect heat balances in the fuel, and 4) the fact that the mechanisms that could lead to heat imbalances have been scrutinized for many years. . . ."

"While there is no way of proving that all possible accident sequences that contribute to public risk have been considered in the study, the systematic approach utilized in identifying possible accident sequences and their dependencies make it very unlikely that a contributor has been overlooked that would significantly change the risk estimate. . . ."

"The principal insights gained in this study are: a. Contrary to the commonly held belief that all nuclear power plant accidents involving core melting would surely result in severe accidents with large public consequences, the magnitudes of the potential consequences of a core melt accident were found to have a wide range of values. The probability is high that the consequences would be modest compared to other types of risks. The likelihood of relatively severe consequences is quite low.

"b. The consequences of reactor accidents are often smaller than many people have believed. Previous AEC studies have been based on unrealistic assumptions and have predicted relatively large consequences for reactors that were much smaller than current reactors. Con-

sequently, there are some who believe, incorrectly, that reactor accidents can produce consequences comparable to that of the explosion of large nuclear weapons. Further, there are many in the nuclear field who have believed that accidents involving melting of the reactor core would always lead to large consequences. This study has shown that predictions of the consequences of nuclear power plant accidents, when performed on a more realistic as opposed to an upper limit basis, are smaller than previous predictions would have led one to believe and, in fact, are no larger, and often smaller, than the consequences of other accidents to which we are already exposed.

"c. The likelihood of reactor accidents is smaller than that of many other accidents having similar consequences. While there are some in the public sector who will feel that the likelihood of occurrence of nuclear power plant accidents should be made essentially zero, neither nuclear accidents nor non-nuclear accidents of any kind can have zero probability. We do not now, and never have, lived in a risk-free world. Nuclear accident risks are relatively low compared to other man-made and natural risks. All other accidents, including fires, explosions, toxic chemical releases, dam failures, earthquakes, hurricanes, and tornadoes, that have been examined in this study are more likely to occur and can have consequences comparable to or greater than nuclear accidents. . . .

"e. The question of what level of risk from nuclear accidents should be accepted by society has not been addressed in this study. It will take consideration by a broader segment of society than that involved in this study to determine what level of nuclear power plant risks should be acceptable. . . ."

Rasmussen study assessed—The Subcommittee on Energy & the Environment of the House Interior & Insular Affairs Committee June 11, 1976 held an oversight hearing to assess the validity of the Rasmussen safety study. At this hearing, the Environmental Protection Agency's Office of Radiation Programs submitted its review of the final report. An abridgement of this review follows:

The EPA found that "several significant

areas . . . [in the study were] either deficient or containing unjustified assumptions. These are (1) failure to fully address
the health effects expected after an accident and to consider adequately a
technical basis for estimating the incidence of the associated bioeffects, which
include a broad range of perspectives, (2)
the assumptions made in regard to
evacuation as a remedial measure, (3)
improperly or incompletely evaluated
parameters used in determining accident
event-sequences and probabilities, and (4)
inadequate description of the analysis of
the consequences of the release of radioactive materials to the environment. . . ."

- The review asserted that "if late so-
matic health effects were adjusted in accordance with EPA's assessment of the
numerical health risks, the estimates
would increase by a factor of from 2 to
10. . . ." The EPA also pointed out the
"deficiencies in the assumptions for
evacuation as a protective action . . .
[which] involves the use of a constant 25-
mile evacuation sector for all core melt
accidents. This is at variance with present
and planned practice and in some cases
overestimates the risk and the need for
evacuation, and in other cases underesti-
mates the risk when larger groups may
have to be evacuated. . . ."

The EPA said that the "report does not
present sufficient information illustrating
the variations in consequences and risk to
nearby population groups caused by
differences in site-specific circumstances,
which may be masked by the restriction of
the analyses of the six composite
sites. . . . The results of our review of the
final report have not altered our opinion
that the Reactor Safety Study provides a
major advance in risk assessment of nuclear power reactors, and that the Study's
general methodology provides a
systematized basis for obtaining useful
assessments of the accident risks where
empirical or historical data are presently
unavailable. . . ."

A critical analysis of the final study
published by Prof. Joel Yellin of MIT in
The Bell Journal of Economics (1976) was
entered into the record of the
congressional hearing.

Yellin questioned the accuracy of the
NRC's estimate that the probability for a

major nuclear accident with serious public
health risks were extremely low. According to Yellin, "[t]he principal conclusion of
the study is that nuclear reactor operation
involves thousands of times less risk to
human life than other industrial activities
and natural processes. This assertion . . .
is entirely based on estimates of the
number of 'early fatalities'—deaths which
would occur within weeks or months of exposure to radioactivity— . . . [and] that,
on the average, each early fatality . . . will
be accompanied by roughly 700 cancer
deaths, 700 cases of genetic defects, 700
spontaneous abortions, and 4,000 cases of
thyroid growths having high incidence
among children. [However,] . . . [n]one of
these large consequences were considered
in making the [study] risk comparisons,
and as a result the conclusion that nuclear
accident risks are relatively very small is
highly misleading. . . ."

Nuclear development endorsed. The continued development of nuclear power as
a commercial energy source was recommended by 34 scientists in a Washington
statement made Jan. 16, 1974 by Dr.
Hans A. Bethe of Cornell University. The
energy crisis, they said, "is the new and
predominant fact of life in industrialized
societies" and there was no currently
viable alternative to increased use of
nuclear power to meet energy needs. "The
U.S. choice is not coal or uranium; we
need both," the statement said. It also de-
plored "the fact that the public is given
unrealistic assurances that there are easy
solutions."

A statement issued the same day by
Ralph Nader's Union of Concerned
Scientists cautioned that "a wide range of
public safety problems must be resolved
before nuclear power plant construction
proceeds in the U.S."

Warnings & Criticism Continue

Warning on hazards. A group of eight
U.S. scientists warned against the

possible dangers of nuclear reactors on public health, the environment and national security during a symposium Nov. 15-16, 1974. The experts suggested at the conclusion of a two-day conference, called Critical Mass 74, that leaders of the U.S. Senate and House investigate the hazards posed by the reactors. The meeting, which was organized by consumer advocate Ralph Nader, drew 650 experts from the U.S., Britain, France and Japan.

A petition drawn up by the eight Americans conceded that scientists had initially welcomed the advent of nuclear power as "a valuable energy source for mankind." But they noted that "this early optimism has been steadily eroded" by the growing realization of the problems posed by relying on commercial nuclear power plants. Among them were the possibility of a catastrophic reactor accident, the chance that some terrorists would steal radioactive materials and build a nuclear bomb and providing safe storage of nuclear wastes for hundreds of years.

AEC denies safety data suppressed. U.S. AEC Chairman Dixy Lee Ray repudiated Nov. 15, 1974 the allegation that the AEC had been suppressing data on the safety of nuclear plants. She conceded that "while there may be some validity for such accusations in the past, the situation has changed today."

Ray's statement was in apparent response to a New York Times report Nov. 10 citing AEC documents showing that since 1964 the commission had sought to conceal studies by its own scientists that found nuclear reactors were more dangerous than officially admitted or raised doubts about safety reactor devices.

Ray cited a number of examples to demonstrate the AEC's openness with the public: the release earlier in 1974 of 25,000 pages of documents developed during the deliberations of the Advisory Committee on Reactor Safety. She also pointed out that in 1973 the AEC had released to the public "an uncompleted 1965 study" that sought to update a 1957 report on the consequences of a possible major accident at a large nuclear power plant. Such a mishap could kill up to 45,-000 persons, and "the possible size of such a disaster might be equal to that of the state of Pennsylvania," the report said.

The cases mentioned by Ray were disputed by Daniel Ford. Ford said that the Advisory Committee released the documents only after being faced with a suit under the Freedom of Information Act, and that many of its pages had been heavily edited. As for the 1973 report, AEC memorandums had shown that the 1965 study had been deliberately withheld from publication by the commission for seven years after its completion and that this study also was finally released under the threat of a Freedom of Information Act suit, Ford said.

The New York Times report said some of the AEC's suppressed documents had been leaked to the Union of Concerned Scientists by commission officials.

Scientists question A-plant program. A petition calling for a "drastic reduction" in the program for construction of nuclear power plants because of safety hazards was sent to the White House and Congress Aug. 6, 1975 by 2,300 scientists.

The petition urged a major research effort on reactor safety, plutonium safeguards and nuclear waste disposal. The "record to date," it said, "evidences many malfunctions of major equipment, operator errors and design defects as well as a continuing weakness in the quality control practices with which nuclear plants are constructed."

On the radioactive waste disposal problem, it said no "technically or economically feasible methods have yet been proven." Nuclear wastes, it added, could prove "a grim legacy . . . to future generations."

A third problem, cited by the petition, involved the potential use of plutonium from commercial reactors for making nuclear weapons. Various studies indicated, it said, "multiple weaknesses in safeguard procedures intended to prevent the theft or diversion of commercial reactor-produced plutonium for use in illicit nuclear explosives or radiological terror weapons."

In addition to calling for a slowdown in the nuclear power plant program, the petition urged a greater national effort to conserve energy and to develop alternate energy sources.

The petition was sponsored by the Union of Concerned Scientists, which circulated it to scientists through mailing lists. The endorsement response was said to have been about 20%.

Atomic engineers resign over safety. Three high-level nuclear engineers for General Electric Co.'s San Jose, Calif. division resigned Feb. 3, 1976 to campaign against nuclear power, a California initiative scheduled for a statewide ballot in June. It would prohibit construction of new nuclear generating plants and require the phasing out of existing plants unable to meet strict conditions. One of the conditions was that the current federal limit of $560 million of liability from any single nuclear accident be eliminated.

The three resigning engineers cited the safety problems as the overriding reason for their move. Aligned with this was their concern that safety problems were not fully reported by nuclear companies to the federal government, which had the regulatory responsibility.

Those resigning, who had from 16 years to 22 years of experience with GE, were Dale G. Bridenbaugh, 44, manager of performance evaluation and improvement at GE's nuclear energy division in San Jose; Richard B. Hubbard, 38, manager of quality assurance; and Gregory C. Minor, 38, manager of advanced control and instrumentation.

A statement from GE Feb. 3 pointed out that the company employed several thousand nuclear engineers and it "emphatically" disagreed with the viewpoint of the three resigning. "The overwhelming majority of the scientific and engineering community, including GE scientists and engineers," it said, "believes the benefits of nuclear power far outweigh the risk."

GE was one of the largest companies in the nuclear-power industry, with orders for 69 U.S. nuclear generating facilities, of which 22 were in operation, and for 48 nu-

clear-power plants overseas, of which 18 were in operation.

In New York state, the federal safety engineer for three nuclear reactors at Indian Point on the Hudson River, 30 miles north of New York City, announced his resignation, citing the safety factor, at a news conference Feb. 9. The engineer, Robert D. Pollard, project manager for the federal Nuclear Regulatory Commission (NRC), planned to become Washington representative of the Union of Concerned Scientists.

Pollard, 36, chief safety engineer for nuclear reactors in North and South Carolina and Texas as well as at Indian Point, said he could not "in conscience remain silent about the perils associated with the United States nuclear-power program." The Indian Point plants, he said, "have been badly designed and constructed and are susceptible to accidents." He recommended closing down Indian Point Plant No. 2 "at once—it's almost an accident waiting to happen."

Consolidated Edison Co., which owned and operated Indian Point Plant No. 2 and one other plant at Indian Point, rejected Pollard's criticism. Spokesmen for the federal Nuclear Regulatory Commission and the State Power Authority, which recently bought the third Indian Point nuclear reactor from Con Ed, also denied that the plants were unsafe.

NRC chairman William A. Anders and Con Ed chairman Charles F. Luce defended the safety of the plants Feb. 10. Anders described Pollard's safety concerns as nothing more than "generic problems." Pollard showed up at the end of Luce's news conference to deliver a letter requesting shutdown of the No. 2 plant.

Delaware River plant barred. The AEC Oct. 5, 1973 denied permission for construction of a plant on a Delaware River island 11 miles from Philadelphia and 4½ miles from Trenton, N.J.

In a letter to Public Service Electric & Gas Co. of New Jersey, the AEC said "population density" near the island was the principle factor in the decision. The AEC suggested an alternate site near Salem, N.J., where two other reactors were already under construction.

Reactor Licensing Problems & Proposed Reforms

A-plant licensing speedup asked. The Carter Administration March 17, 1978 presented legislation that would, the Administration claimed, eventually reduce the time it took to license and build an atomic power plant from the current 10–12 years to about six and one-half years.

Most of the time savings projected by the Administration would come from expediting the licensing process. Currently, approval of construction of a plant and approval of operation of the plant required two separate licensing processes. The bill would combine the two.

The bill would allow sites for nuclear power plants to be selected in advance of specific proposals to build a plant. States would thus be able to "bank" a number of sites for future development.

The bill would limit reconsideration of issues that had already been dealt with in earlier hearings. Under the bill, only state agencies (rather than state agencies and the Nuclear Regulatory Commission) would determine whether a site was acceptable and if there was a need for a new power plant.

The Administration also proposed that atomic reactor designs be standardized. Standardization would allow advance approval.

The bill also provided for government subsidies to parties, such as environmentalists, that wished to take part in the licensing hearings.

The Administration reckoned that the bill, if enacted, would eventually shorten the time for licensing from the approximately four years taken at present to one year. The effects of the bill would not be immediate, because it would take states some time to build up banks of approved sites for nuclear plants.

Construction time would be shortened to five and one-half years from seven years, it was estimated, through the use of standardized reactor designs.

Energy Secretary James Schlesinger, at a news conference March 17, commented, "At the present time, the nuclear option is barely alive." To make nuclear power a real option, Schlesinger said, "we must have reform of the nuclear licensing process."

Sharp Criticism—Spokesmen for some environmental groups sharply criticized the proposal. A statement by the National Resources Defense Council said, "The announcement today that President Carter is sending Jim Schlesinger's version of nuclear licensing reform legislation to Congress represents the final corruption of the President's moral and political courage on the nuclear issue."

Critical Mass Energy Project, a Ralph Nader organization, said the bill would "discourage public participation in regulatory hearings and cause deep potential harm" to the environment.

Critical Mass further claimed that delays with nuclear power plants stemmed chiefly from labor difficulties and problems with equipment, not licensing requirements.

Congressional hearings were later set for the proposed legislation.

Court Action

As the controversy continued, questions over siting suitability, waste disposal and other environmental factors were raised.

Court backs N.Y.C. reactor license. Despite arguments by Justice William O. Douglas that insufficient safety standards had been established, the Supreme Court June 10, 1974 let stand a decision upholding the issuance by the Atomic Energy Commission of a license to build a nuclear reactor on the campus of Columbia University in New York City.

Community groups opposing the research reactor had asked for a review of a May 1972 opinion by the AEC's Atomic Safety and Licensing Appeals Board that the reactor would not be "inimical to the health and safety of the public."

High Court authorizes urban area reactor. The Supreme Court Nov. 11, 1975 overturned a court of appeals decision barring the Nuclear Regulatory Commission from approving a site for a reactor near Portage, Ind. on the south shore of Lake Michigan. The NRC could put a nuclear power plant closer than two miles from an urban area, the high court said, if the plant would be at least two miles from concentrated sites of population within the area. The U.S. 7th Circuit Court of Appeals had set aside the NRC's approval of the site on the ground that the NRC and its predecessor, the Atomic Energy Commission, had violated their own rules governing distance of plants from population centers, defined as 25,000 or more people. The population of Portage was projected to exceed that by the time the plant was scheduled for operation, and the court noted that a state park abutting the site had up to 87,000 visitors a weekend. The suit had been brought by a unit of the Izaak Walton League and a business civic group.

The Court of Appeals had also been critical of the old AEC's policy for "clustering" nuclear plants on the Lake Michigan shore within relatively short distances of Chicago; eight were within 75 miles of downtown Chicago; six more planned, in addition to the Portage site.

The Supreme Court, however, did order the appellate court to hear other arguments against the plant.

Judges curbed on A-plant policy. The Supreme Court April 3, 1978 ruled that federal courts did not have the authority to circumvent government regulations on the building of nuclear power plants. Specifically, the high court backed safety procedures issued by the Nuclear Regulatory Commission in two cases: *Vermont Yankee Nuclear Power Co. v. Natural Resources Defense Council* and *Consumers Power Co. v. Aeschliman.*

The Supreme Court reversed an appeals court decision which was blocking construction of two nuclear power plants, one in Vernon, Vt., the other in Midland, Mich. Justice William H. Rehnquist speaking from the bench, called the lower court rulings an example of "judicial intervention run riot." Writing for the court, Rehnquist said that Congress had established "a reasonable review process [for nuclear power plants] in which courts are to play only a limited role."

In cases of judicial review of agency proceedings, Rehnquist maintained that a court was not free to "stray beyond the judicial providence to explore the procedural format or to impose upon the agency its own notion of which procedures are best or more likely to further some vague, undefined public good."

Justices Harry A. Blackmun and Lewis F. Powell Jr. did not participate in the case.

The ruling was regarded as a major defeat for environmental groups in their battle to limit the construction of atomic reactors.

In both cases, environmental groups had challenged federal safety regulations under which the plants were being built. The U.S. Court of Appeals for Washington, D.C. ruled July 21, 1976 in favor of the groups in both cases. The appeals court invalidated the licenses granted by the NRC to the power companies to build the plants. The court contended that the NRC did not give adequate consideration to the energy-conservation issue, nuclear-waste disposal or to fuel reprocessing.

In writing the unanimous opinions of two three-member appellate panels, U.S. Circuit Chief Judge David L. Bazelon noted the "apprehensive" public concern about nuclear power.

Bazelon held that energy conservation "will have an important . . . role in overall energy policy in coming decades" and that the commission could not dismiss it "without inquiry or explanation" in its licensing process.

The court ordered the commission to hold full licensing hearings on both cases.

A-plant licensing moratorium—The NRC announced Aug. 13 a moratorium on licensing of new powerplants until completion of a study of the environmental hazards of fuel reprocessing and disposal of radioactive wastes. A moratorium also was called against any full-power operation of new A-plants.

The action was in compliance with recent court orders requiring the agency to strengthen its licensing procedures concerning reprocessing and waste disposal.

The commission had issued a staff environmental study Oct. 13 and had proposed at the same time an interim rule to incorporate the environmental impact into its licensing procedures. Licenses issued before adoption of the final rule would be on a conditional basis.

The NRC then announced Nov. 5 that it would resume licensing new nuclear power plants on a conditional basis.

State & Local Action

Calif. rejects nuclear power curbs. In one of the most bitterly-fought contests, California voters June 8, 1976 rejected Proposition 15 to curb nuclear power development. The proposed curb lost by a 2-1 margin—3,986,770–1,924,304. Campaign spending for and against the proposition was estimated at nearly $4 million.

Proposition 15 called for full assumption by utilities of financial liability for nuclear accidents. It would have barred new plant construction and cut back operating plants to 60% of full power output and phased them out unless a two-thirds majority of the state legislature approved plant safety.

The controversy over the issue prompted the legislature to enact "alternative" measures that would postpone new construction pending safety studies of waste disposal practices and storage and recycling of spent atomic fuel. Existing nuclear plants were exempted.

Brown, who took a neutral stance on Proposition 15, signed the measure June 3.

California plant halted. The Pacific Gas & Electric Co. (PG&E) said it was withdrawing license applications submitted to the AEC and the California Public Utilities Commission for construction of an $800 million, 2 million kilowatt nuclear power plant in coastal Mendocino County, because of "unresolved

geological and seismological questions" and "uncertainties" caused by approval of a state coastline protection measure in November 1972, it was reported Jan. 22, 1973.

PG&E said the U.S. Geological Survey had informed the company Jan. 8 that current knowledge of offshore geophysical characteristics was not adequate to resolve questions of site suitability.

N.Y. utilities drop power plans. Consolidated Edison Co. of New York City said July 2, 1973 it had abandoned its plan to build a nuclear power plant on an island in Long Island Sound a half mile from suburban New Rochelle.

The company had bought the island from New Rochelle in 1967 but had not applied for an AEC license. A company spokesman said AEC opposition to reactors near urban areas had been a crucial factor in the decision to drop the plan.

California county votes down A-plant. Residents of Kern County, Calif. March 7 1978 voted by more than a 2-1 margin against the construction of a proposed large nuclear power plant.

The vote—47,282 ballots against the project and 20,591 for it—came on an advisory referendum. While the referendum result was not binding, it was expected to strongly influence the Kern County Board of Supervisors. The board, which had to decide whether to grant a land-use permit for the project, had requested the referendum.

While the issue of nuclear safety entered the debate, opposition to the plant was apparently stimulated more by fears that water required for cooling the atomic reactors would cut into supplies of irrigation water. Kern County, located north of Los Angeles, was a rich agricultural area whose productivity was largely dependent on the availability of irrigation water. The 1977 drought in California had led to water rationing for farmers, heightening awareness of the water issue.

The proposed power plant—called the San Joaquin Valley Nuclear Project—was originally planned to be the largest nuclear power station in the U.S. It would

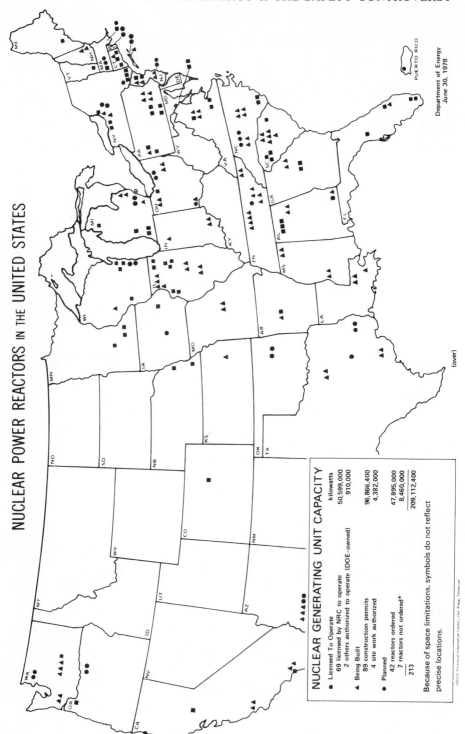

NUCLEAR POWER REACTORS IN THE UNITED STATES

Department of Energy
June 30, 1978

NUCLEAR GENERATING UNIT CAPACITY

		kilowatts
■	Licensed To Operate	
	69 licensed by NRC to operate	50,599,000
	2 others authorized to operate (DOE-owned)	910,000
▲	Being Built	
	89 construction permits	96,866,400
	4 site work authorized	4,382,000
●	Planned	
	42 reactors ordered	47,895,000
	7 reactors not ordered*	8,460,000
	213	209,112,400

Because of space limitations, symbols do not reflect precise locations.

(over)

have had four 1,300-megawatt generating units—enough to provide electricity to a city of five million people. (In response to concern about the water issue, sponsors of the project had tentatively proposed to cut back the plant to only two units.)

The proposed site was 10 miles (16 kilometers) west of Wasco and about 100 miles (160 kilometers) northwest of Los Angeles.

The principal sponsors of the project were the Los Angeles Department of Water and Power and the Pacific Gas and Electric Co. The two sponsors issued a statement after the vote saying they were "disappointed" and "must now reevaluate our position" on the choice of a site.

California Gov. Edmund G. Brown Jr. had opposed the atomic plant. He argued that the state's power needs in the future could be provided through conventional power plants and alternative energy sources, such as solar and wind power and the burning of waste-fiber as fuel.

Large Calif. A-plant voted down. A committee in the Calif. State Assembly April 13, 1978 voted against legislation that would have exempted the proposed $3-billion Sundesert nuclear power plant from a 1976 California law barring the licensing of new nuclear power plants until the federal government, had developed a national system for disposing of spent nuclear fuel. The committee vote apparently meant that the plant would not be built.

The proposed Sundesert site, near Blythe on the Colorado River, had received safety clearances from both the state and federal governments. It had been planned to produce enough electricity for one million to two million people, depending on what size was finally decided upon.

California Gov. Edmund G. Brown Jr. (D) had opposed the project. He argued that the future power needs of the area could be met by nonnuclear sources.

Brown commented that the "legislature would send me that bill [exempting the Sundesert plant from the waste disposal law] in a minute if I gave the slightest signal I would sign it," the Washington Post reported April 14. He added, "I'm never going to give that signal."

The vote against the nuclear project came in the Resources, Land Use and Energy Committee of the State Assembly. The 9–4 vote largely followed party lines: only one Democrat voted to exempt the plant and only one Republican voted against the exemption. The State Senate had already voted to exempt the plant from the 1976 law.

The San Diego Gas and Electric Co. had been the chief backer of the Sundesert project. Even if the project had been exempted from the 1976 law, it would still have faced difficulties in regard to financing. The California Public Utilities Commission reportedly had been unwilling to grant San Diego Gas and Electric a rate increase that the utility had sought to finance construction of the project.

Wisc. A-plant moratorium. The Public Service Commission of Wisconsin ordered a moratorium Aug. 17, 1978 on the construction of new nuclear power plants. The action made Wisconsin the fourth state to curb nuclear power, joining California, Iowa and Maine.

The commission said the moratorium was ordered mainly for "economic reasons." The commission cited the uncertainty of prices for nuclear fuel, the cost of decommissioning nuclear power plants that were too old to continue functioning but were radioactive and the problem of disposing of nuclear wastes.

Charles Cicchetti, chairman of the commission, said the ban was motivated by "a desire to avoid economic catastrophe if the federal government continues to promote nuclear power with unnecessary siting laws, unrealistically strong endorsements and its own incredible inaction."

Cicchetti added, "I am pleased that Wisconsin now joins . . . in laying the nuclear burden upon Washington, where it has belonged for three decades."

Two proposed plants were exempted from the order. But Cicchetti said he thought that various regulatory requirements applying to one of the plants were "nearly insurmountable." The other plant also had "several major regulatory steps to overcome," Cicchetti said, adding that he thought the plant's sponsors would alter their plans.

Wisconsin Gov. Martin J. Schreiber (D) praised the commission's action, saying that nuclear power represented a "billion dollar gamble" for the state.

Anti-Nuclear Protests

Nuclear protests held. Opponents of nuclear energy staged protest demonstrations in South Carolina and Colorado April 29-May 1, 1978. In Portsmouth, N.H., some of those who had taken part in the 1977 protest against the Seabrook nuclear plant gathered April 29 for an anniversary dance at the National Guard armory that was used to house those arrested in the protest.

The South Carolina protest took place at Barnwell, the site of a private plant for reprocessing spent nuclear fuel. The plant, operated by Allied-General Nuclear Services, had not been licensed by the federal government and was not being used for the purpose for which it was designed.

The protest was sponsored by the Palmetto Alliance. Brett Bursey, leader of the organization, said May 1: "The problem is nuclear power. The focus is nuclear waste and the target, albeit a symbolic one, is Allied-General."

Officials of Allied-General had issued a press release April 30 saying, "The owners of the Barnwell plant no longer contemplate its operation as a private commercial reprocessing plant."

On May 1, the third day of the protest gathering, about 250 demonstrators were arrested after marching onto the grounds of the Allied General plant.

The Colorado demonstration focused on the Rocky Flats nuclear weapons facility in Golden. The plant manufactured the plutonium triggers used in hydrogen bombs.

There were about 5,000 protesters, but no arrests, it was reported May 1.

Washington Protest—A demonstration at the construction site of two nuclear power plants near Elma, Washington, resulted in the arrest of 156 persons June 25.

West Coast Protests—Hundreds of nuclear power opponents were arrested Aug. 6 and 7 in demonstrations against the Trojan nuclear plant near Rainier, Ore., and the Diablo Canyon facility near San Luis Obispo in California. The demonstrations, together with others around the country, commemorated the 33rd anniversary of the atomic bombing of Hiroshima.

About 140 demonstrators were arrested at the Trojan power plant, which was owned by Portland General Electric Co. The demonstrators had climbed over a fence to enter the plant grounds.

The plant had been temporarily closed by the Nuclear Regulatory Commission to allow improvements that would strengthen it against earthquakes. The demonstration was planned by the Trojan Decommissioning Alliance, which sought to have the plant permanently closed.

The demonstration against the Diablo Canyon plant was attended by a crowd estimated variously at about 2,000 or nearly 3,000. Nearly 500 persons who illegally entered the grounds of the plant were arrested, law enforcement authorities said.

The protest was organized by a coalition of environmental, peace and other groups that called itself the Abalone Alliance.

The Pacific Gas and Electric Co. owned the plant, which had been completed in 1976 but not put into service because of questions about its ability to withstand earthquakes.

The Diablo Canyon facility was designed to stand up to earthquake shocks measuring slightly over six on the Richter scale. During construction, geologists discovered a fault line three miles offshore near the plant. Some experts believed that the fault could give rise to earthquakes considerably more powerful than the plant was designed to withstand.

Seabrook Dispute

After five years of costly legal delays caused by dispute over safety and environ-

mental issues, the Nuclear Regulatory Commission (NRC) announced Aug. 10, 1978 that construction would resume on the controversial nuclear power plant in Seabrook, New Hampshire. The announcement followed the Environmental Protection Agency (EPA) ruling Aug. 4, that the proposed open-ocean cooling system would comply with federal water pollution laws.

The NRC had first halted construction in March 1977 because of EPA charges that the cooling system posed a danger to marine life. Following EPA approval in June 1977, the Public Service Co. of New Hampshire resumed construction. However, in June 1978 the NRC again ordered a halt to construction pending the completion of an EPA study of the cooling system. The NRC also announced that it would review a possible alternative site for the nuclear facility.

To environmentalists and other opponents of nuclear energy who fought to block its construction, the Seabrook plant had come to symbolize the controversy over nuclear power in the U.S. During 1976-78, the construction site was the scene of several major protests in which more than 1,000 people were arrested.

Seabrook protests. About 2,000 demonstrators moved onto the construction site of a nuclear power generating plant in Seabrook, N.H. April 30, 1977. Equipped with tents, food and medical supplies, they vowed to "occupy" the site until plans to build the plant were abandoned.

When the protesters refused to leave May 1, about 300 state police officers from New Hampshire and neighboring states began making arrests. The demonstrators did not resist. State officials said 1,414 people had been arrested through early May 2 on charges of criminal trespass.

The well-organized and nonviolent demonstration was planned by a group that called itself the Clamshell Alliance. The alliance was made up largely of members of existing groups who had grown frustrated with what they felt was the predisposition of the federal Nuclear Regulatory Commission (NRC) to favor construction of nuclear power plants.

The proposed $2-billion, twin-tower plant in Seabrook had been embroiled in controversy for the previous four years, as environmentalists and other nuclear energy opponents fought and delayed it in the federal courts. The plant had become a symbol of the widening national debate over whether nuclear energy should be used extensively as a power source in the U.S.

The Clamshell Alliance had held a demonstration at the Seabrook plant in August 1976 at which 180 of 1,000 protestors had been arrested.

Most of those arrested May 1-2 refused bail in an effort to have everyone released on a bond of personal recognizance. According to figures compiled in the demonstrators' headquarters, 631 people were being held May 3 in a National Guard armory in Manchester, 267 in Dover, 195 in Somersworth and 220 in Concord, along with two smaller groups in county jails. The Rockingham County prosecutor said May 2 that "several" persons had pleaded guilty and been sentenced to 120 days in jail.

The state authorities had agreed to release only New Hampshire residents without cash bail. The state position had been upheld May 4 by the New Hampshire Supreme Court.

The more than 600 demonstrators still imprisoned were released from their National Guard armory prisons May 13. Many of those arrested had already been freed on bail.

Under an agreement with the Rockingham County prosecutor, the demonstrators pleaded not guilty to criminal trespass charges in county district court. They were then found guilty in mass trials, sentenced to 15-day jail terms and $100 fines and released on their own recognizance pending their automatic appeals to county superior court. As part of the agreement, the Clamshell Alliance, which had negotiated for the protestors, said it would help the part-time county judicial staff with paperwork and scheduling of trials.

(Under the two-tiered county court system in New Hampshire, defendants convicted in district court had the right of automatic appeal to superior court. The district courts existed mainly to deal with traffic violations and other minor and un-contested cases.)

Seventeen protestors had been tried in-dividually in district court May 5. Others had been tried and convicted some days later. In the first cases, a judge had sentenced each of the demonstrators to serve 15 days at hard labor and pay a $100 fine. In a break with local legal practice, he had ordered that the sentences be served immediately, before appeal to su-perior court.

The judge had suspended the sentence of the first demonstrator convicted but had reversed himself after an appearance in court by state Attorney General David Souter. Souter said a suspended sentence was like no sentence at all. Asking for harsher penalties, he called the Seabrook occupation "one of the most well-planned acts of criminal activity" in the nation's history and said the state police had over-heard citizens' band radio messages indi-cating that the demonstrators planned to reoccupy the construction site.

The total cost of confining the demon-strators was approximately $50,000 a day, state officials said. The National Guard bill alone was $290,566. That figure did not include costs of state or local police, county sheriffs or court facilities.

The U.S. Law Enforcement Assistance Administration (LEAA) had turned down a request by New Hampshire for $699,461 to pay the Seabrook case costs, it was reported May 19. The LEAA had said it would not pay National Guard costs un-less the Guard was on federal duty.

In an earlier appeal for aid, Gov. Mel-drim Thomson May 6 had called on "cor-porations, labor unions and rank-and-file citizens throughout America" to help New Hampshire pay demonstration-related costs. "Our battle of today can become theirs of tomorrow," Thomson said. The Seabrook occupation was the first massive show of civil disobedience in the nation in opposition to construction of a nuclear power plant.

If completed, the 2,300-megawatt plant would provide 80% of New Hampshire's electricity by the mid-1980s, while selling half its power to other New England states.

A-power backers rally—Several thou-sand supporters of nuclear power rallied in Manchester, New Hampshire, June 26. Employes of the Long Island Lighting Company had been offered free bus trips by their company if they wished to attend the rally; construction workers and in-dustry representatives were also present in force. (A spokesman for the Long Is-land utility said the company put "absolutely no pressure" on employes to attend.)

Speakers at the rally depicted nuclear energy as essential to jobs and economic vitality. They criticized opponents of nu-clear power as proponents of a "no-growth philosophy."

Construction suspended 4 months. The construction of the Seabrook power plant by the Public Service Co. of New Hampshire had been halted by the NRC March 31, 1977 because of "uncertainty" over the outcome of a pend-ing U.S. Environmental Protection Agency (EPA) review of plans for the plant's cooling system. The EPA regional office had objected to use of a cooling system that would suck in about 1.2 billion gallons of sea water a day and return it to the ocean 38 degrees warmer, which would be dangerous to seashore and marine life. The NRC action confirmed a Jan. 21 decision by its Atomic Safety and Licensing Board Panel suspending construction permits for the Seabrook facility.

Power companies, a number of business interests and construction unions in New Hampshire and Massachusetts contended that continued difficulties at the long-delayed plant were another example of government red tape. They said the EPA objections showed an unwarranted concern with fish rather than with produc-ing energy.

The antinuclear forces charged that atomic power in general, and the Seabrook plant in particular, was a dan-gerous and ill-conceived venture in which faulty construction or maintenance could lead to a disastrous accident. They also objected to the possible danger of cancer from low-level radiation escaping from the plant.

The decision of the regional EPA ad-ministrator in Boston, under which the

NRC had halted the plant's construction, was reversed by EPA Administrator Douglas Costle June 17. Costle said that the EPA would approve the plant's water cooling system. But approval by the NRC was still required before the Public Service Co. could resume construction.

After the NRC's Atomic Safety and Licensing Board again approved the permit, construction was started in August 1977. Three of the four NRC commissioners later affirmed the Licensing Board decision Jan. 7, 1978.

In announcing approval of the proposed cooling system, Costle stressed that construction of the Seabrook plant raised a number of issues but that most of them were "outside the scope of this decision." He added, "This decision isn't a go or no-go signal on nuclear power. It isn't a seal of environmental approval on the Seabrook plant."

"My only authority," Costle said, "was to determine whether the record of past hearings on Seabrook showed that the heated water discharged by the plant wouldn't harm ocean life and that the proposed cooling system reflects the best technology for minimizing adverse environmental impacts. I think the record shows that."

Although approving the cooling system, Costle added two conditions. Water for discharge had to be piped 7,000 feet out into the ocean, not 3,000 feet as proposed by Public Service Co. (The water further from shore was less densely populated with sea life.) Also, Costle ruled that backflushing a process involving the discharge of water hotter than usually discharged could only be done when the tide was going out and so would carry the hot water quickly out into the ocean.

Two oceanographers hired by the EPA to independently analyze the environmental impact of the Seabrook plant criticized Costle's decision. Both Edward Carpenter, professor at the Stony Brook campus of the State University of New York, and Theodore Smayda, professor at the University of Rhode Island, said the available data was inadequate to conclude the Seabrook plant would not harm sea life.

Later, environmentalists charged that the EPA had violated required procedures in granting the approval of the cooling system: specifically, and environmentalists said that a group of technical experts that was supposed only to review evidence concerning the cooling system had actually submitted new evidence.

A federal appeals court in Boston ruled in the environmentalists' favor in February. The court told the EPA to discard the review panel's findings or allow the panel to be cross-examined by the opponents of the plant.

The EPA then agreed to reopen hearings on the cooling system to allow cross-examination of the panel and also to permit new evidence to be introduced.

Review of Seabrook site backed. The NRC's Atomic Safety and Licensing Appeal Board April 28, 1978 ordered a new study conducted to find the best site for the nucelar plant being constructed at Seabrook.

The Appeal Board ruled, 2–1, that a more extensive search for a plant site should have been made. The Appeal Board instructed a lower NRC licensing board to examine other possible sites in both northern and southern New England.

The Appeal Board refused, however to cancel the preliminary construction licenses granted for the Seabrook site.

The board's action came in response to an appeal filed by the state of Massachusetts, the New England Coalition on Nuclear Pollution, the Seacoast Antipollution League and the Audubon Society of New Hampshire.

A-Backers Demonstrate—Over 1,000 construction workers rallied May 10 at the statehouse in Concord, New Hampshire to show support for the Seabrook nuclear power plant. The demonstration was organized by the state Building and Construction Trades Council.

Protests resume. Opponents of nuclear power gathered in a three-day demonstration June 24–26, 1978 against the Seabrook plant in particular and nuclear power in general.

In contrast with the protest at this site in the spring of 1977, civil disobedience tactics were not used and the demonstration ended peacefully without the mass arrests of the 1977 protest.

The Clamshell Alliance sponsored the demonstration. Before the protest began, the Alliance made an agreement with the state government and the Public Service Co. of New Hampshire on ground rules for the demonstration.

The agreement gave the protesters an 18-acre plot in which to rally, camp out and hold an alternative energy fair. In exchange, the protesters agreed to leave the site after three days and not attempt to illegally occupy the area where construction on the plant was actually going on.

The demonstration reached its peak June 25, when there was a crowd variously estimated at 10,000 to 20,000. On the next day, about 2,000 protesters staged a march in Manchester, the largest city in New Hampshire. The Federal Nuclear Regulatory Commission and the Environmental Protection Agency were holding hearings there on the Seabrook plant.

Attending the Seabrook protest were a number of celebrities associated with the anti-war movement of the 1960s: Benjamin Spock, the pediatrician; comedian Dick Gregory, and singers Pete Seeger and Arlo Guthrie.

The demonstration was hailed as a success by Harvey Wasserman, one of the leaders of the Clamshell Alliance. "This demonstration," Wasserman said June 26, "brought out the closet clams [that is, supporters of the Alliance's goal of stopping the nuclear power plant] .... I've always thought we'd stop this plant, but at times I've been weary; but now I'm sure."

New Hampshire Gov. Meldrim Thomson Jr., an avid supporter of the Seabrook plant, claimed June 26 that the demonstration was a failure. "At no time was there one minute of construction time lost by the workers," Thomson said, "and at no time was any portion of construction halted as a result of either direct or indirect actions by the demonstrators. The Clamshell Alliance therefore has experienced what must be to them a very distinct and humiliating defeat."

Thomson attended a rally June 25 in Manchester staged by supporters of the Seabrook plant.

2nd halt ordered at Seabrook. The NRC, by a 2-1 vote June 30, 1978, ordered the suspension of construction of the Seabrook power plant for the second time in two years.

The indefinite delay took effect July 21.

Concern over two issues prompted the commission to order the halt. The NRC said that there had not been an adequate effort to determine if there might be alternative sites preferable to the Seabrook location. The commission ordered an examination of possible sites in northern New England (previously, only sites in southern New England had been considered).

Also, the NRC noted that the Environmental Protection Agency was conducting a study of the Seabrook facility's cooling system. The commission held that it would be wise to delay construction until the study was completed. In an alternative cooling system under consideration coolant water would be recycled by being pumped up high towers where it would cool for reuse.)

Construction at Seabrook had progressed to the point that the plant was about 10% complete. The NRC ruling observed that "dropping the site comparison . . . merely on the basis that events have advanced too far would mean that no matter what errors are committed, no matter what warnings have been received, if enough work is done on the site quickly enough the facility is an accomplished fact, whether the National Environmental Policy Act has been complied with or not." That would be "unacceptable," the NRC said.

The order to stop work was hailed by opponents of the plant. A group called the Seabrook Natural Guard termed the ruling "an historic breakthrough for the entire anti-nuclear movement in this country."

Gov. Thomson called the ruling "asinine" and threatened to bring suit against the federal government for damages.

Thomson said the ruling, if allowed to stand, would "wreak havoc on New Hampshire's economy" and "also have a severe rippling effect on the economy of the rest of New England."

The construction halt would result in the layoff of about 1,800 of the 2,200 construction workers at the plant. Public Service Co. asseted that the cost of suspending construction would be about $500,000 a day.

Cooling system approved, construction resumes. The EPA ruled Aug. 4, 1978 that the proposed open-ocean cooling system planned for the Seabrook power plant would comply with federal water pollution laws.

The EPA ruling was followed by a Nuclear Regulatory Commission decision Aug. 10 clearing the way for resumption of construction of the facility.

EPA Administrator Douglas Costle said in his ruling that the discharged water would result in "a 5-degree maximum surface temperature rise." The discharged coolant, which would be carried out into the ocean through 2.5 mile tunnels (4.2 kilometers), "would not have a significant effect" on the "population of fish, shellfish and wildlife in and on the receiving waters." Costle. said. Costle emphasized that his decision dealt only with the plant's cooling system. "I haven't considered," he said, "nor may I consider in the context of these proceedings, whether construction of the Seabrook plant is desirable from an overall environmental perspective."

Opponents of the Seabrook facility were dismayed by the decision. The Clamshell Alliance, said that the ruling was "a serious, even tragic, violation of the responsibilities given the EPA by the American people."

A spokeswoman for the coalition said that if construction was allowed to start again on the Seabrook plant, "citizens around the nation will be forced to conclude we have no meaningful recourse through the regulatory agencies."

Backers of the nuclear plant hailed Costle's decision. Gov. Thomson said, "All of our sky is bright and sunny today. Every cloud has a silver lining."

NRC OKs Construction—The Nuclear Regulatory Commission Aug. 10 ruled, 4-0, that construction of the Seabrook facility could resume immediately.

The commission said that Costle's decision "eliminates the condition which led to the suspension of the Seabrook construction permits."

A review of possible alternative sites for the nuclear facility was still pending, but commission staff members said that was not considered a strong enough reason to continue the construction halt. Officials said the site review was seen largely as a contingency measure in case the EPA approval of the cooling system was nullified by the courts and a different system for which the Seabrook site was not appropriate was required.

Public Service Co. said the halt had cost $12 million.

Seabrook Foes Vow Action—The Clamshell Alliance responded to the NRC decision by saying it planned a series of small "nonviolent, civil disobedience" protests at Seabrook over the coming months.

The Breeder Challenge & Other A-Fuel Issues

In a report on "Atomic Energy Programs" released in 1973, the AEC said that the U.S.' "primary reactor research and development effort ... [would be] directed toward [the] successful development and utility acceptance of the breeder reactor, so-called because it 'breeds' more fuel than it consumes. Development emphasis [would be] on the liquid metal-cooled fast breeder reactor. There also [would be] developmental efforts for the light-water breeder reactor." During the 1950s and 1960s, nuclear powerplant development concentrated on nonbreeder reactors—the light-water and gas-cooled types.

The AEC said: "The breeder reactor is essential to realization of the full potential of nuclear power. During the course of their 30-year operating lives, the water reactors offer sufficient economic advantages to overcome the immediate capital cost advantages of coal and oil powerplants. However, water reactors are relatively inefficient thermally, and use only about 2% of the heat energy potential

*of the uranium fuel. In addition, water
reactors are dependent upon the
availability of large amounts of low-cost
uranium ore.... The liquid metal fast
breeder reactor (LMFBR), however, uti-
lizes 60 to 90 percent of the uranium
processed from ore and will extend ore
reserves significantly. The breeder's more
efficient fuel usage could make the cost of
producing electricity relatively insensitive
to the cost of uranium.... [T]he LMFBR
will have greater thermal efficiency than
the current water reactors; thus, much
less waste heat will be discharged into the
environment. The LMFBR will also
provide a premium market for the plutonium being produced as a byproduct by
water reactors and will ... produce additional plutonium.... Five other nations—the United Kingdom, France,
Japan, The Federal Republic of Germany,
and the Soviet Union—also have work underway on breeder reactors...."*

Breeder-reactor project supported. A
General Accounting Office report to Congress July 30, 1975 supported continued
development of the controversial fast-
breeder nuclear reactor. It said the
country was perhaps seven to 10 years
away from the time it would have to make
"a firm decision" whether to commit itself to the liquid metal fast breeder
reactor as a basic central station energy
source. In the meantime, it recommended
the government continue its $10.7 billion
project to develop the reactor. It also
recommended top priority be given to development of better safeguards for the
highly radioactive wastes that would be
produced by the breeders.

The major safety concerns related to
the commercial light water reactor
(LWR) differ from those of an FBR in that
removal of coolant or movement of fuel
tends to slow down the neutron reaction in
the LWR, whereas the reverse may be
true in an FBR. On the other hand, because an LWR operates at high pressure,
a leak may lead to a rapid loss of coolant
and a requirement for emergency core
cooling, while in an FBR, which operates
at low pressure, the safety systems may
continue to cool the core.

ERDA Administrator Robert C.
Seamans Jr. announced Jan. 2, 1976 a decision to proceed with a 10-year research
and development program for a liquid
metal fast breeder reactor.

Among the unresolved issues of the
program were reactor safety and protection and waste management, major issues
of the current reactor program. According to ERDA deputy administrator Robert
Fri. Jan. 2, the question was "do we stop
the program altogether or continue it to resolve the problems. We've got to conduct
the program to answer the questions."

Court curbs plutonium use. The Court
of Appeals in New York May 26, 1976
barred commercial use of plutonium as a
nuclear fuel until the Nuclear Regulatory
Commission completed a safety study.
The ruling reversed a decision by the commission in November, 1975 to grant interim licenses for the manufacture and
use of plutonium pending a decision on
permanent licensing.

The NRC ruling had been challenged by
environmental organizations, including the
Natural Resources Defense Council, and
the state of New York. The NRC was
supported by 28 companies involved in nuclear power, including the Babcock &
Wilcox Co. and Westinghouse Electric
Corp.

The court, basing its opinion on the National Environmental Policy Act, found
that a draft study on use of plutonium was
inadequate in considering both alternative
power sources and the problems of theft
or sabotage of plutonium.

The AEC had recommended in August
1974 that plutonium be used to supplement uranium as a fuel in boiling-water
reactors as a way to reduce the cost of
producing electric power.

Some experts, such as Dr. Theodore
Taylor, a former AEC scientist,
contended that widespread use of plutonium could be dangerous because it could
be converted without much sophisticated
training into homemade weapons.

The theory was put to the test by television producer John Angier, who asked a
20-year old undergraduate student at
Massachusetts Institute of Technology to

see if he could develop a nuclear bomb in his spare time without talking to experts and using only publicly available reference works. He was given five weeks for the assignment. His plan was submitted for evaluation to a nuclear scientist at the Swedish defense ministry, who concluded, after consulting several colleagues, that there was a "fair chance" the device could explode, with a yield range of from 100,-000 to two million pounds of TNT.

The student reported that designing and building the bomb, assuming one had the plutonium, was not much harder "than building a motorcycle." He thought the bomb could be made for $10,000 to $30,-000. It would be about the size of a large desk, weigh 550–1,000 pounds and require 11–19 pounds of plutonium.

A copy of his plan was sent to the AEC. The only other copy was destroyed, along with the notes of the student, who had demanded anonymity.

The test gained national prominence by broadcast on public television in March as part of a scientific series called Nova.

Breeder safety study proposed. The EPA April 27, 1976 recommended further study of the safety and radioactive-waste disposal problems of a plutonium-breeding nuclear reactor. While it was not "necessarily advocating a delay" in development of such a reactor, it said, "sufficient evidence exists to warrant re-examination" of the timing of the breeder program. The suggestion in the report was for a delay of from four to 12 years.

The EPA was critical of some AEC projections—of the probable growth rate of electric power demand, which the EPA found overstated; of the chance of a major accident being one in 10 million, which the EPA called premature; and of the time and effort needed for adequate resolution of environmental and safety problems, which the EPA deemed "highly optimistic."

In 1974 the agency had asked that a new accident survey be completed.

Ford proposes private A-fuel plants. President Ford asked Congress June 26, 1975 to open the nuclear fuel field to private industry. The government had had exclusive operation in that area since the birth of the atomic bomb.

Private industry would be permitted to build and operate new uranium enrichment ·plants, where uranium was processed into fuel for atomic plants producing electrical power. The government would sell its enrichment technology for royalties. The government would step back in on projects that failed because of regulatory or financial problems, the private participants would be paid off and the facility built by the government. Congress was to have veto power over construction plans.

The proposal contained provisions to guarantee the secrecy of the enrichment technology and to subject exports to safeguards and controls. The plants built in the U.S. by private enterprise would be controlled and regulated by the federal government.

Much of the world's enriched uranium currently was produced in three U.S. plants constructed originally for the weapons program. They were operated by private contractors for the Energy Research and Development Administration (ERDA). The plants were operating at capacity and no new orders had been taken since 1974. The Administration proposed to retain government ownership of these three plants.

The President told Congress more nuclear fuel plants were urgently needed "if we are to insure an adequate supply of enriched uranium for the nuclear power needs of the future and if we are to retain our position as a major supplier of enriched uranium to the world."

He said his proposal would result in "enormous savings to the American taxpayer." Other arguments for private entry into the field were that the market for nuclear fuel was "predominantly in the private sector" and the process of uranium enrichment was "clearly industrial in nature."

ERDA had already received various proposals from private industry for building such plants.

Administration plan scored—The Ford Administration's proposal to bring private industry into the nuclear power field was criticized by the General Accounting Office Nov. 1. In a report, requested by Sen. John O. Pastore (D, R.I.), chairman of the Joint Committee on Atomic Energy, the GAO contended that a preferable way to increase use of enriched uranium in the nuclear power field would be to expand existing gaseous diffusion plants rather than have private industry build a new facility.

The GAO report was focused largely on a proposal of Uranium Enrichment Associates, a partnership of Bechtel Corp. and Goodyear Tire & Rubber Co., for a $2.75 billion gas-diffusion process project.

The GAO considered the Bechtel-Goodyear project more costly than adding on to current plants and probably subject to more project delay. "Its fundamental shortcoming," however, the GAO said, was that the industry plan assured a good profit to private investors while "shifting most of the risk during construction and proving the plant can operate to the government."

It considered the Bechtel-Goodyear project "not acceptable."

Other projects proposed in response to President Ford's plan were a $700 million project by Exxon Corp., a $900 million plant by a Signal Cos. subsidiary and a $1 billion joint venture by Atlantic Richfield Co. and Electro-Nucleonics Corp., the Wall Street Journal reported Oct. 2. All these would employ the gas-centrifuge process.

Ford proposes tighter nuclear curbs— President Ford proposed tighter control over nuclear power Oct. 28, 1976 in an effort to curb the worldwide spread of nuclear weapons. The new policy was announced at the White House and by Ford in Cincinnati, where he was campaigning. Ford noted there that the new policy included an earlier proposal to Congress for a $4.4-billion addition to a federal uranium fuel plant at Portsmouth, Ohio, which, Ford said, would mean "6,000 new jobs for southern Ohio."

A statement Oct. 28 from the headquarters of Jimmy Carter, then the Democratic presidential nominee, characterized Ford's new policy as "a shortsighted, campaign-inspired attempt to correct the timid record of the past."

Carter's statement said that Ford's plan "only thinly disguises" his interest in a massive aid program for plutonium reprocessing "on a so-called evaluation basis."

Under Ford's new policy, domestic reprocessing of nuclear fuel would no longer be considered "inevitable" and would not be undertaken "unless there is sound reason to conclude that the world community can effectively overcome the associated risks of [weapons] proliferation."

The new policy involved a shift from a large-scale "demonstration program" in this area to an "evaluation program" taking into account the weapons risk. The President called upon other nuclear nations to forsake, for at least three years, the spread of technology or facilities for reprocessing.

Also in the international area, Ford proposed:

■ To offer "binding letters of intent" to assure countries that the U.S. would supply them with nuclear fuel if they accepted "responsible restraints" on their nuclear policies.

■ To initiate diplomatic negotiations to strengthen international nuclear controls.

■ To revise U.S. bilateral pacts toward anti-proliferation policy.

■ To establish an international authority to store spent nuclear fuel of various nations and to accept, in the meantime and under certain conditions, the spent fuel of other nations for safekeeping.

Carter opposes plutonium use. After the election of Jimmy Carter and his inauguration as president, White House energy adviser James R. Schlesinger said March 25, 1977 that the Carter Administration was opposed to the development of plutonium fuel systems for nuclear power plants. "For the immediate future we will not be using plutonium recycling," Schlesinger said.

He addressed his remarks to 19 men and women participating in a White House energy conference. The 19 had been selected from among 20,000 respondents to a questionnaire on energy policy sent out by the White House March 2 to approximately 450,000 persons, 300,000 of whom had been chosen at random. The other 150,000 recipients included members of Congress and business and civic leaders.

Schlesinger indicated that the Administration would go along with construction of more conventional uranium-fueled reactors to help meet the country's electricity needs over the next 25 years. He said the Administration wanted to separate conventional reactors "from the plutonium economy" and "separate the use of nuclear power from the spread of nuclear weapons."

The nuclear-power industry and the Ford Administration had contended that existing uranium-fueled nuclear plants would need a new fuel by the end of the 20th century. They had argued that the nation's uranium resources would run so low that plutonium use would become necessary. They added that the plutonium-fueled fast-breeder reactor currently being developed would produce more plutonium than it consumed and thus would be an inexhaustible energy source.

But Carter Administration officials and some academic experts believed the nation's uranium resources had been underestimated and there was enough uranium to fuel existing types of plants well into the 21st century.

Breeder reactor, plutonium use scored. A panel of 21 scientists and economists March 21, 1977 recommended changes in the nuclear energy policy of the U.S., including an end to the crash program to develop a commercial fast-breeder reactor and an indefinite postponement of plans to reprocess plutonium for use as a reactor fuel.

The panel, organized by the Mitre Corp. with a grant from the Ford Foundation, endorsed the continued use of current generation uranium-fueled nuclear power plants, but urged more emphasis on solving safety and radioactive-waste problems. The panelists said adequate electricity for the short term could be generated by a combination of nuclear and coal-fired plants. As for the 21st century, they said, "We believe that some mix of

coal, solar and fusion energy, assisted by conservation, would be capable of supplying society's long-term energy needs."

The panelists, agreeing with critics of the breeder reactor and plutonium reprocessing, objected primarily to the danger of the worldwide spread of nuclear weapons as a result of the availability of plutonium. They said federal research on the breeder should be continued, but in such a way that commercial use of the breeder would be delayed. (The federal government in recent years had poured billions of dollars into breeder-reactor research.) They said further that plutonium reprocessing had "little if any" economic value and should be "postponed indefinitely."

The report was based on two key assumptions: that conventional nuclear reactors, which currently generated about 10% of the electricity in the U.S., would be about as expensive and dangerous to operate as coal-fired plants; and that the current official estimates of uranium reserves and resources substantially underestimated the amount of uranium that would be available during the next two or three decades.

The Ford Foundation said the panelists, who had worked on the study for almost two years, had been chosen because of their middle-of-the-road views on nuclear energy. Among them were John Sawhill, former head of the Federal Energy Administration, and two members of the Carter Administration: Defense Secretary Harold Brown, who was president of the California Institute of Technology when the work on the report was done, and Joseph S. Nye Jr., deputy to the undersecretary of security assistance in the State Department.

Plutonium fuel system called safe—The Ford-Mitre study was strongly criticized by two nuclear researchers at the Electric Power Research Institute in Palo Alto, Calif., in interviews reported March 26. Milton Levenson and Chauncy Starr, officials of the institute, noted that a breeder reactor nuclear fuel-recycling system in which the fuel remained so radioactive that terrorists could not have stolen it had operated almost unnoticed for more than four years in the 1960s at what was currently known as the Idaho National Engineering Laboratory.

Starr said "major flaws" in the Ford report included "unverified and risky" assumptions about future uranium and coal supplies and prices and a failure to estimate the "economic and social penalties" if energy supplies turned out to be smaller than forecast.

Both men attacked the study's implicit assumption that fuel-reprocessing for future breeder reactors would be done the same way as at existing plants for recovering plutonium from spent uranium reactor fuel. Such plants dissolved the spent fuel completely so that near-pure plutonium could be separated from radioactive by-products of the nuclear reaction.

Levenson said that in a fast-breeder reactor, such radioactive debris would not interfere with the nuclear reaction and so the plutonium fuel would not have to be pure. The fuel could remain highly radioactive and thus, he said, immune to "diversion" by terrorists.

Levenson said the process used in Idaho had been developed not to guard against terrorism but to speed up the fuel-reprocessing operation.

Carter defers plutonium use. President Carter proposed April 7, 1977 that the U.S. "defer indefinitely" the use of plutonium as a fuel in commercial nuclear power plants. Carter also indicated the government's fast-breeder reactor program would be slowed and reshaped to emphasize the development of fuels other than plutonium.

The President said concern about the spread of nuclear weapons internationally had been the main factor in his decision to defer the use of plutonium. In a statement intended as a signal to other nations he said, "We have concluded that a viable and economic nuclear power program can be sustained" without the reprocessing of plutonium.

In line with his general proposal, the President said a plant under construction at Barnwell, S.C. would "receive neither federal encouragement nor funding for its completion as a reprocessing facility." Allied-General Nuclear Services had already spent an estimated $250 million on

the plant and had sought as much as $500 million in federal assistance to complete it as a fuel-reprocessing demonstration project. Allied-General was indirectly equally owned by Allied Chemical Corp. and General Atomic Co., itself a joint venture of Gulf Oil Corp. and the Royal/Dutch Shell Group.

Carter said his Administration would "restructure the U.S. breeder-reactor program to give greater priority to alternative designs . . . and to defer the date when breeder reactors will be put into commercial use." Under questioning from reporters, he said the government's fast-breeder demonstration project on the Clinch River near Oak Ridge, Tenn. would "not be terminated as such" but would receive a reduced amount of federal funds. Future government support of the facility, he said, would focus on experimental research and development rather than the possibility of commercial utilization.

Knowledgeable sources said that top environmental officials in the Administration had recommended that the Clinch River project be discontinued altogether.

The U.S. Energy Research and Development Administration (ERDA) estimated that $500 million had been spent on designing the Clinch River reactor and that additional outlays of $50 million to $100 million would be required just to complete the design effort.

Carter said that to make sure the U.S. did not need to use plutonium, he would propose an increase in the production of enriched uranium for nuclear power plant use. He also said he would propose legislation to allow the U.S. to guarantee supplies of uranium fuel to foreign countries so they would not have to rely on reprocessed plutonium either. (According to the Washington Post April 13, the U.S. was insisting that countries that purchased nuclear power plants from the U.S. waive their rights to the plutonium generated by burning U.S.-supplied uranium in those plants.

The President said the U.S. would continue to embargo the export of technology or equipment for uranium enrichment or plutonium reprocessing. (Both processes could be used to make nuclear weapons.) He said the U.S. would continue discussions aimed at reducing the spread of atomic weapons with countries that supplied and used nuclear materials. But he said he would not ask countries that had plutonium-reprocessing capabilities, such as West Germany, France, Great Britain and Japan, to discontinue reprocessing.

"We are not trying to impose our will," the President said, noting that other industrial powers had a "special need" that the U.S. did not have for atomic energy because they lacked the domestic supplies of coal, oil and natural gas that the U.S. had. "But I hope," Carter added, "that by this unilateral action we can set a standard."

He also said he hoped those countries with reprocessing capability would help prevent other nations from acquiring it. That issue was due to be considered at a nonproliferation conference of industrial powers in London in late April and again at the London economic summit of Western countries in early May.

Reaction in U.S. mixed, cautious abroad—In the U.S., President Carter's statement drew criticism from the Atomic Industrial Forum, a trade group financed by reactor manufacturers and others with business interests in the nuclear energy field. A group spokesman said Carter was "in effect asking Americans to forego an essential element in their energy future in order to create a bargaining chip" for use in international negotiations. Other U.S. energy industry officials expressed relief that the President had not decided to completely cancel research and development of the breeder reactor.

A coalition of 12 environmental and other groups, including the Sierra Club and the National Council of Churches, lauded Carter's ban on plutonium use but said his "failure to take an equally strong and definitive position on the breeder-reactor program may seriously undermine the effect of this as a nonproliferation policy."

Abroad, the statement met generally polite but firm resistance. It was pointed out in many capitals that Carter had not made an outright declaration that U.S.-supplied fuels could not be reprocessed. Great Britain, France, the U.S.S.R.,

West Germany and Japan had all invested heavily in plutonium-fueled fast-breeder reactors as insurance against exhaustion of uranium supplies or dependence on foreign uranium suppliers. Moreover, Britain and France had been negotiating large fuel-reprocessing contracts with Japan and other nations. (The precise implications of Carter's statement were especially important to Japan, which received all its enriched uranium from the U.S.)

Most of the nuclear power stations in the world used a U.S.-designed reactor that needed enriched uranium fuel—uranium with a 93% saturation of the isotope Uranium-235. Enriched uranium supplied by the U.S. was exported under special license from the federal Nuclear Regulatory Commission in compliance with the terms of the Nuclear Nonproliferation Treaty. Even after use, the nuclear material—which would then contain plutonium—could not be transferred to another country for storage or reprocessing without another license that complied with arms control regulations.

The Carter Administration, however, currently was holding up approval of 28 export licenses involving the shipment of enriched uranium fuel to 13 countries, including most of the nations of Western Europe, Japan, Brazil and Canada, according to a Washington Post report April 14. The Post April 13 had reported that in the case of Spain, the U.S. had said it would approve certain export licenses if Washington were given a veto over when and where spent fuel was reprocessed, if at all. The Post said the U.S. was asking for veto power over all spent fuel, no matter who had supplied it.

The French government April 8 noted that Carter had said he would not impose the U.S. will on other countries, which the French took to mean that he did not intend to prejudge the means other countries might use eventually to meet their energy needs. In that respect, French observers noted, Carter had taken into account the comments presented by France, West Germany, Great Britain and Japan during U.S.-initiated talks held prior to the President's speech.

The West German government declined to comment on the statement. Hours before Carter spoke, West Germany had issued its own nuclear-policy statement calling for curtailment of the spread of nuclear weapons through "multinational, nondiscriminatory and generally binding" agreements on safeguards and the peaceful uses of nuclear energy rather than through the restrictions on technology advocated by Carter. The German statement pointed out that the peaceful use of nuclear energy was for many countries necessary to secure their social and economic progress.

New enriched-uranium contracts. Energy Secretary James Schlesinger announced at a news conference May 26, 1978 that the U.S. would "reopen the order books" for long-term contracts to provide enriched uranium.

The U.S. had suspended long-term contracting for enriched uranium on June 30, 1974, because the capacity of the government's three gaseous diffusion (enrichment) plants had been fully obligated. New contracts could now be signed because the government was acquiring additional enrichment capacity. Also, fewer nuclear power plants were being built or planned than originally had been predicted, so the projected demand for enriched uranium had declined.

(Enrichment was the process by which, starting with a mixture of U-238 and U-235, the percentage of U-235 in the mixture was increased to the point where the uranium could be used to fuel nuclear power plants. U-235 was the fissionable isotope of uranium.)

Schlesinger said that reopening the order books would allow the Energy Department to "fulfill President Carter's commitment to make the United States a reliable supplier of enriched uranium to the world."

Carter's national energy program. President Carter presented a national energy program to Congress April 20, 1977. A major goal in the program was the conversion of industry and utilities using oil and natural gas to coal and other more abundant fuels. Carter had told the nation in a televised address April 18 that even with the conversion effort, there would be a gap between need and availability of energy. "Therefore, as a last resort we must continue to use increasing amounts of nuclear energy."

He said the U.S. had 63 nuclear power plants, producing about 3% of the country's total energy, and about 70 more plants were licensed for construction.

Carter said there was "no need to enter the plutonium age" by licensing or building a fast-breeder reactor such as the proposed demonstration plant on the Clinch River in Tennessee.

He called, however, for an increase in the capacity to produce enriched uranium fuel for light-water nuclear power plants using new centrifuge technology.

A reform of licensing procedures was urged to upgrade the safety of the plants and accelerate decisions.

Among details of his nuclear power program*:

—There would be definite deferral of commercial reprocessing and recycling of spent fuels produced in U.S. civilian nuclear power plants. (administrative)

—Indefinite deferral of construction of the Clinch River liquid-metal fast-breeder reactor demonstration project in Tennessee. The federal breeder-reactor program would be re-oriented toward research and development of non-plutonium fuels, with emphasis on safety and nonproliferation of plutonium technology. (administrative)

—Expansion of the U.S. uranium-enrichment capacity, with use of energy-saving centrifuge plants rather than the gaseous diffusion plants currently in use. (administrative/budget)

—Legislation to guarantee sales of enriched uranium to any country that agreed to comply with U.S. nonproliferation

---

*Source: The White House. The designation in parentheses following each item in the energy program indicates whether the proposal would have to be implemented administratively, by congressional action, through the regular federal budget process, or through a combination of means.

objectives and certain other conditions. (legislative)

—Expansion of the Nuclear Regulatory Commission (NRC) staff; increase in the number of nuclear plant inspections by the NRC; development by the NRC of siting criteria for new nuclear plants; mandatory rather than voluntary reporting to the NRC of all minor mishaps and component failures at nuclear plants, and review of the NRC licensing process for nuclear plants to insure its objectivity and efficiency. (administrative)

—Review of the federal nuclear waste-disposal program. (administrative)

Carter energy plan generally praised— President Carter's efforts toward the establishment of a national energy policy were applauded by industrialists and environmentalists, businesses and foreign governments and others throughout the country and the world; however, almost all found one or more specific proposals to criticize.

As could be expected, the nature of the criticism varied according to the particular concern of the critic:

Foreign—Despite concern in many quarters over President Carter's proposals on nuclear energy, foreign governments and international agencies reacted generally favorably to Carter's energy policy statement.

Officials of the European Community (EC) Commission hailed the program as "important and courageous" April 21 and expressed the hope that Carter's message would inspire the EC to take similar action.

In Paris, an International Energy Agency spokesman April 21 called the program "well balanced."

Clinch River project debate. Construction of the Clinch River breeder reactor in Oak Ridge, Tenn., a 350-megawatt demonstration liquid metal fast breeder (LMFB), was first authorized by President Nixon in 1970. The objectives of the project were to demonstrate the LMFB's reliability for electric power

generation and to establish a strong competitive breeder industry by the mid-1980s. However, after his election as president, Jimmy Carter opposed the uranium-plutonium fueled breeder as unneeded, uneconomical and a threat to the spread of nuclear weapons. Congress, however, strongly supported the project, seeing in it the answer to the nation's energy needs.

Carter signs Clinch River appropriations bill, vows to terminate project. President Carter March 7, 1978 signed a bill providing $7.8 billion in supplemental appropriations for the 1978 fiscal year. The bill included an $80-million appropriation for further development of the fast breeder reactor at Clinch River, Tenn. Carter, who had opposed the project, had vetoed a 1977 bill authorizing progress on the fast breeder. In signing the 1978 bill, Carter said the project was "neither necessary to meet our projected energy needs nor economically sound." Carter said, however, that he would use some of the $80 million to terminate the project.

The President had noted that the estimated cost of the Clinch River project had grown to more than $2.2 billion from the estimate of $450 million when it had been first authorized in 1970.

"I am committed to a vigorous energy research and development strategy to ensure maximum progress on shifting the energy base of the United States away from oil and natural gas," Carter said. "However, I am also concerned about the risk of introducing the plutonium economy through an unnecessary commercial demonstration facility."

Construction of the Clinch River plant, he said, "in no way is necessary to ensure the continued development of nuclear technologies, including liquid metal fast breeder technology."

Breeder reactor halt called illegal— Elmer Staats, comptroller general of the U.S.,* told President Carter, Energy

---

*As comptroller general of the U.S., Staats headed the General Accounting Office, which was the accounting and investigative arm of Congress. The comptroller general was empowered to determine personal liability for the misuse of federal funds.

Secretary James Schlesinger and other Administration officials March 10 that it would be illegal to halt development of the Clinch River breeder reactor.

Staats said that it was the opinion of the GAO that the legislation had been written so as to require "that the funds available for this project be used only for the design, development, construction and operation of the liquid-metal fast-breeder reactor and that they may not be used to terminate such activities."

Any Administration official who ordered the funds spent to close down the project would be held personally liable for the money spent, Staats said.

Breeder reactor compromise offered. With Congress and the Administration at odds over the Clinch River breeder reactor, the Energy Department had prepared a compromise proposal by which the Clinch River project would be terminated, but additional funding would be devoted to alternative breeder designs. The proposal, described in the New York Times March 17 and the Wall Street Journal March 23, was apparently the product of negotiations between Energy Secretary James Schlesinger and Rep. Walter Flowers (D, Ala.).

(Flowers was chairman of the subcommittee on fossil and nuclear energy research, development and demonstration of the House Science and Technology Committee.)

The compromise proposal involved a two-year study to investigate alternatives to the Clinch River technology. Schlesinger, in a letter to House Science and Technology Committee Chairman Olin Teague (D, Tex.), said, "The Administration believes the breeder program should be reoriented to evaluate designs for a larger advanced fission facility."

The study would look at breeder reactors designed to produce power in the range of 650 to 900 megawatts, which would be double to triple the capacity of the Clinch River project. The uranium-thorium fuel cycle, considered safer than the uranium-plutonium cycle, would presumably be studied, as would new

technologies considered to be more secure from the point of view of weapons proliferation.

One such method—called Civex—had been announced by U.S. and British researchers Feb. 27. In the Civex process spent fuel would be treated so that it could be reused as fuel. But the plutonium in it would not at any stage be purified to the extent that it could be used for a bomb.

The scientists also said that the fuel, at every stage of the process, would be so highly radioactive that it could not be handled directly by human beings, a fact that would presumably deter terrorists from attempting to steal the material.

The Civex process could be commercially developed in about 10 years, the scientists said.

The compromise proposal did not, however, include a definite commitment to construct a new breeder reactor. The decision on construction would be deferred until the study was completed. Some members of Congress indicated they could not accept a compromise on the Clinch River issue that lacked a commitment to construction. Rep. Mike McCormack (D, Wash.), quoted in the New York Times March 17, said the proposal as reported was "not satisfactory." "Why in the world," he continued, "should Congress give up everything we've fought for?"

Study backs breeder effort. A study sponsored by the Rockefeller Foundation and issued May 10, 1978 said the U.S. should pursue research on plutonium-fueled breeder reactors in cooperation with Japan and possibly Great Britain. The report, titled *International Cooperation on Breeder Reactors,* was prepared by International Energy Associates Limited, a group of private researchers in Washington.

The Rockefeller Foundation study took issue with the anti-proliferation argument. The study asserted that U.S. restraint on breeders would have little impact, because other countries would go ahead with programs to develop plutonium-fueled breeders.

John Gray, main author of the report, said at a news conference May 10 that the U.S. was "being left behind not only in the evolution of breeder technology, but also in its ability to constructively influence how the world decides to use them."

Mason Willrich, an official at the Rockefeller Foundation, noted that breeder reactors were designed to consume plutonium. Willrich said, "If you think through the consequences" of using current-day atomic reactors, but not breeder reactors, "the result is the accumulation of large amounts of plutonium spread around the world."

The study noted that there were alternative types of breeder technology under investigation that would not use a plutonium fuel cycle. But the study favored development of the plutonium-fueled breeders, because the weight of technological experience was with that type. Gray said it was the "only system that can be ready for use by the end of this century."

The study advocated partnership with Japan and possibly Great Britain because both countries had done considerable research in breeder technology and did not have partners at present. Germany, France, Belgium, Italy and the Netherlands were already committed to working with each other, the study observed.

Possible drawbacks to the breeder reactors were the extreme combustibility of liquid sodium used in the process, and the dangerously long radioactive life of plutonium, the by-product fuel.

Plutonium defended. In an effort to investigate every aspect of the Clinch River breeder reactor, Rep. Olin E. Teague (D, Tex.) entered an article into the Congressional Record July 15, 1977 which was written by Professor Bernard L. Cohen, a nuclear engineer. In his article, Cohen condemned the "insidious campaign of slander and half-truths that has blackened the public image of plutonium." He said:·

One line of attack has been on the toxicity of plutonium. It is frequently called "the most toxic substance known to man". Not so! Botulism toxin, for example, is a thousand times more toxic. A widely read and quoted book states that plutonium-239 is "1000 times more toxic than modern nerve gases". Again not so! Per ounce inhaled, the two are about comparably toxic; but plutonium

is a solid material which can only be harmful if inhaled in the form of a very fine dust, so mechanisms are required to suspend it in air. Given the choice of being in a room with an open one pound can of plutonium or nerve gas, the latter would be infinitely more dangerous. A recent favorite statement in newspapers and magazines is that "a single particle of plutonium inhaled into the lung will cause lung cancer". Absolutely false! The largest particle that can get past the body's defense mechanisms and enter the lung is about 5 microns (0.0002 inches) in diameter and it would take about a million plutonium particles of this size to give a 50-percent risk of lung cancer. ...

Cohen said that "25 men who worked with plutonium in the 1940s [had] inhaled millions of plutonium particles into their lungs . . . , and none of them has yet had lung cancer."

The article went on to discuss the problems faced in the design and fabrication of a bomb from plutonium by would be terrorists: "This requires reasonable expertise in nuclear reactor physics, chemistry, electronics, machining and mechanical operations, and high explosives technology, plus some familiarity with health physics and miscellaneous areas. There was a television program about an MIT student who designed a bomb that had a small chance of exploding (this latter point was not mentioned), but he only did the design on paper which involves only the first of the skills listed above; it is by no means the most difficult part of the task. The work would require thousands of dollars worth of equipment and would take many weeks or months of concerted effort. There would be substantial danger, perhaps a 30 percent risk of fatal injury, and the bomb would be far from certain of working. . . ."

Plutonium tested in humans in 1940s— The Energy Research and Development Administration (ERDA) had confirmed that the government injected plutonium into human subjects from 1945 to 1947 to determine what the substance would do to workers manufacturing the bomb, The Washington Post reported Feb. 22, 1976.

The injections were administered in varying quantities to eighteen persons (13 males and 5 females) ranging in age from four to 68. All were believed suffering from terminal illnesses.

Officials of the ERDA said records on the experiment were so unclear that it was not known how the subjects were selected or who ordered the tests. The ERDA could only be certain that one person had consented to the injection.

According to a fact sheet prepared by the ERDA, the purpose of the experiment was to gather information that would permit scientists to determine the amount of plutonium to which humans could safely be exposed.

The fact sheet said that seven persons who received injections lived less than one year afterward; three lived between one and three years; two between 14 and 20 years; and one 28 years. The fate of two was unknown, and three were still living.

COMMERCIAL NUCLEAR POWER REACTORS IN THE UNITED STATES

SITE	PLANT NAME	CAPACITY NET kW(e)	UTILITY	COMMERCIAL OPERATION
ALABAMA				
Decatur	Browns Ferry Nuclear Power Plant: Unit 1	1,065,000	Tennessee Valley Authority	1974
Decatur	Browns Ferry Nuclear Power Plant: Unit 2	1,065,000	Tennessee Valley Authority	1975
Decatur	Browns Ferry Nuclear Power Plant: Unit 3	1,065,000	Tennessee Valley Authority	1977
Dothan	Joseph M. Farley Nuclear Plant: Unit 1	829,000	Alabama Power Co.	1977
Dothan	Joseph M. Farley Nuclear Plant: Unit 2	820,000	Alabama Power Co.	1980
Scottsboro	Bellefonte Nuclear Plant: Unit 1	1,213,000	Tennessee Valley Authority	1980
Scottsboro	Bellefonte Nuclear Plant: Unit 2	1,213,000	Tennessee Valley Authority	1981
ARIZONA				
Wintersburg	Palo Verde Nuclear Generating Station: Unit 1	1,270,700	Arizona Public Service	1983
Wintersburg	Palo Verde Nuclear Generating Station: Unit 2	1,270,700	Arizona Public Service	1984
Wintersburg	Palo Verde Nuclear Generating Station: Unit 3	1,270,700	Arizona Public Service	1986
Wintersburg	Palo Verde Nuclear Generating Station: Unit 4	1,270,700	Arizona Public Service	1988
Wintersburg	Palo Verde Nuclear Generating Station: Unit 5	1,270,000	Arizona Public Service	1990
ARKANSAS				
Russellville	Arkansas Nuclear One: Unit 1	850,000	Arkansas Power & Light Co.	1974
Russellville	Arkansas Nuclear One: Unit 2	912,000	Arkansas Power & Light Co.	1978
CALIFORNIA				
Eureka	Humboldt Bay Power Plant: Unit 3	63,000	Pacific Gas & Electric Co.	1963
San Clemente	San Onofre Nuclear Generating Station: Unit 1	436,000	So. Calif. Ed. & San Diego Gas & El. Co.	1968
San Clemente	San Onofre Nuclear Generating Station: Unit 2	1,100,000	So. Calif. Ed. & San Diego Gas & El. Co.	1981
San Clemente	San Onofre Nuclear Generating Station: Unit 3	1,100,000	So. Calif. Ed. & San Diego Gas & El. Co.	1983
Diablo Canyon	Diablo Canyon Nuclear Power Plant: Unit 1	1,084,000	Pacific Gas & Electric Co.	1978
Diablo Canyon	Diablo Canyon Nuclear Power Plant: Unit 2	1,106,000	Pacific Gas & Electric Co.	1979
Clay Station	Rancho Seco Nuclear Generating Station	918,000	Sacramento Municipal Utility District	1975
Site not selected	Unit 1	1,200,000	Pacific Gas & Electric Co.	Indef.
Site not selected	Unit 2	1,200,000	Pacific Gas & Electric Co.	Indef.
COLORADO				
Platteville	Ft. St. Vrain Nuclear Generating Station	330,000	Public Service Co. of Colorado	1978
CONNECTICUT				
Haddam Neck	Haddam Neck Plant	575,000	Conn. Yankee Atomic Power Co.	1968
Waterford	Millstone Nuclear Power Station: Unit 1	660,000	Northeast Nuclear Energy Co.	1971
Waterford	Millstone Nuclear Power Station: Unit 2	830,000	Northeast Nuclear Energy Co.	1975
Waterford	Millstone Nuclear Power Station: Unit 3	1,156,000	Northeast Nuclear Energy Co.	1986
FLORIDA				
Florida City	Turkey Point Station: Unit 3	693,000	Florida Power & Light Co.	1972
Florida City	Turkey Point Station: Unit 4	693,000	Florida Power & Light Co.	1973
Red Level	Crystal River Plant: Unit 3	825,000	Florida Power Corp.	1977
Ft. Pierce	St. Lucie Plant: Unit 1	802,000	Florida Power & Light Co.	1976
Ft. Pierce	St. Lucie Plant: Unit 2	810,000	Florida Power & Light Co.	1983
GEORGIA				
Baxley	Edwin I. Hatch Nuclear Plant: Unit 1	786,000	Georgia Power Co.	1975
Baxley	Edwin I. Hatch Nuclear Plant: Unit 2	795,000	Georgia Power Co.	1978
Waynesboro	Alvin W. Vogtle, Jr. Plant: Unit 1	1,110,000	Georgia Power Co.	1984
Waynesboro	Alvin W. Vogtle, Jr. Plant: Unit 2	1,110,000	Georgia Power Co.	1985
ILLINOIS				
Morris	Dresden Nuclear Power Station: Unit 1	200,000	Commonwealth Edison Co.	1960
Morris	Dresden Nuclear Power Station: Unit 2	794,000	Commonwealth Edison Co.	1970
Morris	Dresden Nuclear Power Station: Unit 3	794,000	Commonwealth Edison Co.	1971
Zion	Zion Nuclear Plant: Unit 1	1,040,000	Commonwealth Edison Co.	1973

Zion	Zion Nuclear Plant: Unit 2	1,040,000	Commonwealth Edison Co.	1974
Cordova	Quad-Cities Station: Unit 1	789,000	Comm. Ed. Co.-Ia.-Ill. Gas & Elec. Co.	1972
Cordova	Quad-Cities Station: Unit 2	789,000	Comm. Ed. Co.-Ia.-Ill. Gas & Elec. Co.	1972
Seneca	LaSalle County Nuclear Station: Unit 1	1,078,000	Commonwealth Edison Co.	1979
Seneca	LaSalle County Nuclear Station: Unit 2	1,078,000	Commonwealth Edison Co.	1980
Byron	Byron Station: Unit 1	1,120,000	Commonwealth Edison Co.	1981
Byron	Byron Station: Unit 2	1,120,000	Commonwealth Edison Co.	1982
Braidwood	Braidwood: Unit 1	1,120,000	Commonwealth Edison Co.	1981
Braidwood	Braidwood: Unit 2	1,120,000	Commonwealth Edison Co.	1982
Clinton	Clinton Nuclear Power Plant: Unit 1	933,400	Illinois Power Co.	1982
Clinton	Clinton Nuclear Power Plant: Unit 2	933,400	Illinois Power Co.	1988
INDIANA				
Westchester	Bailly Generating Station	645,800	Northern Indiana Public Service Co.	1984
Madison	Marble Hill Nuclear Power Station: Unit 1	1,130,000	Public Service Indiana	1982
Madison	Marble Hill Nuclear Power Station: Unit 2	1,130,000	Public Service Indiana	1984
IOWA				
Palo	Duane Arnold Energy Center: Unit 1	538,000	Iowa Electric Light and Power Co.	1975
Vandalia	Vandalia Nuclear Project	1,270,000	Iowa Power & Light Co.	Indef.
KANSAS				
Burlington	Wolf Creek Generating Station: Unit 1	1,150,000	Kansas Gas & Electric—Kansas City P&L	1983
LOUISIANA				
Taft	Waterford Generating Station: Unit 3	1,113,000	Louisiana Power & Light Co.	1981
St. Francisville	River Bend Station: Unit 1	934,000	Gulf States Utilities Co.	1983
St. Francisville	River Bend Station: Unit 2	934,000	Gulf States Utilities Co.	1985
MAINE				
Wiscasset	Maine Yankee Atomic Power Plant	790,000	Maine Yankee Atomic Power Co.	1972
MARYLAND				
Lusby	Calvert Cliffs Nuclear Power Plant: Unit 1	845,000	Baltimore Gas and Electric Co.	1975
Lusby	Calvert Cliffs Nuclear Power Plant: Unit 2	845,000	Baltimore Gas and Electric Co.	1977
MASSACHUSETTS				
Rowe	Yankee Nuclear Power Station	175,000	Yankee Atomic Electric Co.	1961
Plymouth	Pilgrim Station: Unit 1	655,000	Boston Edison Co.	1972
Plymouth	Pilgrim Station: Unit 2	1,150,000	Boston Edison Co.	1985
Montague	Montague: Unit 1	1,150,000	Northeast Utilities	1988
Montague	Montague: Unit 2	1,150,000	Northeast Utilities	1990
MICHIGAN				
Big Rock Point	Big Rock Point Nuclear Plant	72,000	Consumers Power Co.	1963
South Haven	Palisades Nuclear Power Station	805,000	Consumers Power Co.	1971
Newport	Enrico Fermi Atomic Power Plant: Unit 2	1,093,000	Detroit Edison Co.	1980
Bridgman	Donald C. Cook Plant: Unit 1	1,054,000	Indiana & Michigan Electric Co.	1975
Bridgman	Donald C. Cook Plant: Unit 2	1,100,000	Indiana & Michigan Electric Co.	1978
Midland	Midland Nuclear Power Plant: Unit 1	460,000	Consumers Power Co.	1982
Midland	Midland Nuclear Power Plant: Unit 2	811,000	Consumers Power Co.	1981
St. Clair County	Greenwood: Unit 2	1,200,000	Detroit Edison Co.	1987
St. Clair County	Greenwood: Unit 3	1,200,000	Detroit Edison Co.	1989
MINNESOTA				
Monticello	Monticello Nuclear Generating Plant	545,000	Northern States Power Co.	1971
Red Wing	Prairie Island Nuclear Generating Plant: Unit 1	530,000	Northern States Power Co.	1973
Red Wing	Prairie Island Nuclear Generating Plant: Unit 2	530,000	Northern States Power Co.	1974
MISSISSIPPI				
Corinth	Yellow Creek: Unit 1	1,285,000	Tennessee Valley Authority	1985
Corinth	Yellow Creek: Unit 2	1,285,000	Tennessee Valley Authority	1986
Port Gibson	Grand Gulf Nuclear Station: Unit 1	1,250,000	Mississippi Power & Light Co.	1981
Port Gibson	Grand Gulf Nuclear Station: Unit 2	1,250,000	Mississippi Power & Light Co.	1984
MISSOURI				
Fulton	Callaway Plant: Unit 1	1,120,000	Union Electric Co.	1983
Fulton	Callaway Plant: Unit 2	1,120,000	Union Electric Co.	1987
NEBRASKA				
Fort Calhoun	Ft. Calhoun Station: Unit 1	457,000	Omaha Public Power District	1973
Brownville	Cooper Nuclear Station	778,000	Nebraska Pub. Pow. Dist. & Iowa P&L Co.	1974
NEW HAMPSHIRE				
Seabrook	Seabrook Nuclear Station: Unit 1	1,200,000	Public Service of N.H.	1982
Seabrook	Seabrook Nuclear Station: Unit 2	1,200,000	Public Service of N.H.	1984
NEW JERSEY				
Toms River	Oyster Creek Nuclear Power Plant: Unit 1	650,000	Jersey Central Power & Light Co.	1969
Forked River	Forked River Generating Station: Unit 1	1,070,000	Jersey Central Power & Light Co.	1983
Salem	Salem Nuclear Generating Station: Unit 1	1,090,000	Public Service Electric and Gas, N.J.	1977
Salem	Salem Nuclear Generating Station: Unit 2	1,115,000	Public Service Electric and Gas, N.J.	1979
Salem	Hope Creek Generating Station: Unit 1	1,067,000	Public Service Electric and Gas, N.J.	1984
Salem	Hope Creek Generating Station: Unit 2	1,067,000	Public Service Electric and Gas, N.J.	1986

SITE	PLANT NAME	CAPACITY NET kW(e)	UTILITY	COMMERCIAL OPERATION
Little Egg Inlet	Atlantic Generating Station: Unit 1	1,150,000	Public Service Electric and Gas, N.J.	1988
Little Egg Inlet	Atlantic Generating Station: Unit 2	1,150,000	Public Service Electric and Gas, N.J.	1990
Site not selected	1993 Unit	1,150,000	Public Service Electric and Gas, N.J.	1993
Site not selected	1995 Unit	1,150,000	Public Service Electric and Gas, N.J.	1995
NEW YORK				
Buchanan	Indian Point Station: Unit 1	265,000	Consolidated Edison Co.	1962
Buchanan	Indian Point Station: Unit 2	873,000	Consolidated Edison Co.	1973
Buchanan	Indian Point Station: Unit 3	965,000	Power Authority of State of N.Y.	1976
Scriba	Nine Mile Point Nuclear Station: Unit 1	610,000	Niagara Mohawk Power Corp.	1969
Scriba	Nine Mile Point Nuclear Station: Unit 2	1,099,800	Niagara Mohawk Power Corp.	1983
Ontario	R.E. Ginna Nuclear Power Plant: Unit 1	490,000	Rochester Gas & Electric Corp.	1970
Brookhaven	Shoreham Nuclear Power Station	819,000	Long Island Lighting Co.	1980
Scriba	James A. FitzPatrick Nuclear Power Plant	821,000	Power Authority of State of N.Y.	1975
Cementon	Greene County Nuclear Power Plant	1,212,000	Power Authority of State of N.Y.	1986
Jamesport	Jamesport 1	1,150,000	Long Island Lighting Co.	1988
Jamesport	Jamesport 2	1,150,000	Long Island Lighting Co.	1990
Oswego	Sterling Nuclear: Unit 1	1,150,000	Rochester Gas & Electric Co.	1986
New Haven	Unit 1	1,250,000	New York State Electric & Gas Co.	1991
New Haven	Unit 2	1,250,000	New York State Electric & Gas Co.	1993
NORTH CAROLINA				
Southport	Brunswick Steam Electric Plant: Unit 1	821,000	Carolina Power and Light Co.	1977
Southport	Brunswick Steam Electric Plant: Unit 2	821,000	Carolina Power and Light Co.	1975
Cowans Ford Dam	Wm. B. McGuire Nuclear Station: Unit 1	1,180,000	Duke Power Co.	1979
Cowans Ford Dam	Wm. B. McGuire Nuclear Station: Unit 2	1,180,000	Duke Power Co.	1981
Bonsal	Shearon Harris Plant: Unit 1	900,000	Carolina Power and Light Co.	1984
Bonsal	Shearon Harris Plant: Unit 2	900,000	Carolina Power and Light Co.	1986
Bonsal	Shearon Harris Plant: Unit 3	900,000	Carolina Power and Light Co.	1990
Bonsal	Shearon Harris Plant: Unit 4	900,000	Carolina Power and Light Co.	1988
Davie County	Perkins Nuclear Station: Unit 1	1,280,000	Duke Power Co.	1988
Davie County	Perkins Nuclear Station: Unit 2	1,280,000	Duke Power Co.	1991
Davie County	Perkins Nuclear Station: Unit 3	1,280,000	Duke Power Co.	1993
Site not selected		1,150,000	Carolina Power and Light Co.	Indef.
Site not selected		1,150,000	Carolina Power and Light Co.	Indef.
OHIO				
Berlin Heights	Erie: Unit 1	1,260,000	Ohio Edison Co.	1986
Berlin Heights	Erie: Unit 2	1,260,000	Ohio Edison Co.	1988
Oak Harbor	Davis-Besse Nuclear Power Station: Unit 1	906,000	Toledo Edison-Cleveland El. Illum. Co.	1977
Oak Harbor	Davis-Besse Nuclear Power Station: Unit 2	906,000	Toledo Edison-Cleveland El. Illum. Co.	1985
Oak Harbor	Davis-Besse Nuclear Power Station: Unit 3	906,000	Toledo Edison-Cleveland El. Illum. Co.	1987
Perry	Perry Nuclear Power Plant: Unit 1	1,205,000	Cleveland Electric Illuminating Co.	1981
Perry	Perry Nuclear Power Plant: Unit 2	1,205,000	Cleveland Electric Illuminating Co.	1983
Moscow	Wm. H. Zimmer Nuclear Power Station: Unit 1	810,000	Cincinnati Gas & Electric Co.	1980
Moscow	Wm. H. Zimmer Nuclear Power Station: Unit 2	1,170,000	Cincinnati Gas & Electric Co.	1989
OKLAHOMA				
Inola	Black Fox Nuclear Station: Unit 1	1,150,000	Public Service of Oklahoma	1984
Inola	Black Fox Nuclear Station: Unit 2	1,150,000	Public Service of Oklahoma	1986
OREGON				
Prescott	Trojan Nuclear Plant: Unit 1	1,130,000	Portland General Electric Co.	1976
Arlington	Pebble Springs Nuclear Plant: Unit 1	1,260,000	Portland General Electric Co.	1986
Arlington	Pebble Springs Nuclear Plant: Unit 2	1,260,000	Portland General Electric Co.	1989
PENNSYLVANIA				
Peach Bottom	Peach Bottom Atomic Power Station: Unit 2	1,065,000	Philadelphia Electric Co.	1974
Peach Bottom	Peach Bottom Atomic Power Station: Unit 3	1,065,000	Philadelphia Electric Co.	1974
Pottstown	Limerick Generating Station: Unit 1	1,065,000	Philadelphia Electric Co.	1983
Pottstown	Limerick Generating Station: Unit 2	1,065,000	Philadelphia Electric Co.	1985
Shippingport	Shippingport Atomic Power Station[3]	60,000	Department of Energy	1957
Shippingport	Beaver Valley Power Station: Unit 1	852,000	Duquesne Light Co.—Ohio Edison Co.	1976
Shippingport	Beaver Valley Power Station: Unit 2	833,000	Duquesne Light Co.—Ohio Edison Co.	1982
Middletown	Three Mile Island Nuclear Station: Unit 1	819,000	Metropolitan Edison Co.	1974
Middletown	Three Mile Island Nuclear Station: Unit 2	906,000	Jersey Central Power & Light Co.	1978
Berwick	Susquehanna Steam Electric Station: Unit 1	1,050,000	Pennsylvania Power and Light	1981
Berwick	Susquehanna Steam Electric Station: Unit 2	1,050,000	Pennsylvania Power and Light	1982
RHODE ISLAND				
Charlestown	New England Power (NEP): Unit 1	1,150,000	New England Power Co.	1986
Charlestown	New England Power (NEP): Unit 2	1,150,000	New England Power Co.	1988
SOUTH CAROLINA				
Hartsville	H.B. Robinson S.E. Plant: Unit 2	700,000	Carolina Power and Light Co.	1971
Seneca	Oconee Nuclear Station: Unit 1	887,000	Duke Power Co.	1973
Seneca	Oconee Nuclear Station: Unit 2	887,000	Duke Power Co.	1974

Seneca	Oconee Nuclear Station: Unit 3	887,000	Duke Power Co.	1974
Broad River	Virgil C. Summer Nuclear Station: Unit 1	900,000	South Carolina Electric and Gas Co.	1980
Lake Wylie	Catawba Nuclear Station: Unit 1	1,145,000	Duke Power Co.	1981
Lake Wylie	Catawba Nuclear Station: Unit 2	1,145,000	Duke Power Co.	1983
Cherokee County	Cherokee Nuclear Station: Unit 1	1,280,000	Duke Power Co.	1985
Cherokee County	Cherokee Nuclear Station: Unit 2	1,280,000	Duke Power Co.	1987
Cherokee County	Cherokee Nuclear Station: Unit 3	1,280,000	Duke Power Co.	1989

TENNESSEE

Daisy	Sequoyah Nuclear Power Plant: Unit 1	1,148,000	Tennessee Valley Authority	1979
Daisy	Sequoyah Nuclear Power Plant: Unit 2	1,148,000	Tennessee Valley Authority	1980
Spring City	Watts Bar Nuclear Plant: Unit 1	1,177,000	Tennessee Valley Authority	1980
Spring City	Watts Bar Nuclear Plant: Unit 2	1,177,000	Tennessee Valley Authority	1981
Oak Ridge	Clinch River Breeder Reactor Plant	350,000	Department of Energy	Indef.
Hartsville	A, Unit 1	1,233,000	Tennessee Valley Authority	1983
Hartsville	A, Unit 2	1,233,000	Tennessee Valley Authority	1984
Hartsville	B, Unit 1	1,233,000	Tennessee Valley Authority	1983
Hartsville	B, Unit 2	1,233,000	Tennessee Valley Authority	1984
Kingsport	Phipps Bend, Unit 1	1,233,000	Tennessee Valley Authority	1984
Kingsport	Phipps Bend, Unit 2	1,233,000	Tennessee Valley Authority	1985

TEXAS

Glen Rose	Comanche Peak Steam Electric Station: Unit 1	1,111,000	Texas Utilities Services, Inc.	1981
Glen Rose	Comanche Peak Steam Electric Station: Unit 2	1,111,000	Texas Utilities Services, Inc.	1983
Jasper	Blue Hills: Unit 1	918,000	Gulf States Utilities	1989
Jasper	Blue Hills: Unit 2	918,000	Gulf States Utilities	1991
Wallis	Allens Creek: Unit 1	1,150,000	Houston Lighting & Power Co.	1985
Matagorda County	South Texas: Unit 1	1,250,000	Central Power & Lt.–Houston Lt. & Power	1980
Matagorda County	South Texas: Unit 2	1,250,000	Central Power & Lt.–Houston Lt. & Power	1982

VERMONT

Vernon	Vermont Yankee Generating Station	514,000	Vermont Yankee Nuclear Power Corp.	1972

VIRGINIA

Gravel Neck	Surry Power Station: Unit 1	822,000	Virginia Electric & Power Company	1972
Gravel Neck	Surry Power Station: Unit 2	822,000	Virginia Electric & Power Company	1973
Mineral	North Anna Power Station: Unit 1	907,000	Virginia Electric & Power Company	1978
Mineral	North Anna Power Station: Unit 2	907,000	Virginia Electric & Power Company	1979
Mineral	North Anna Power Station: Unit 3	907,000	Virginia Electric & Power Company	1983
Mineral	North Anna Power Station: Unit 4	907,000	Virginia Electric & Power Company	1984

WASHINGTON

Richland	N-Reactor/WPPSS Steam	850,000	Department of Energy	1966
Richland	WPPSS No. 1	1,218,000	Washington Public Power Supply System	1982
Richland	WPPSS No. 2	1,100,000	Washington Public Power Supply System	1980
Satsop	WPPSS No. 3	1,242,000	Washington Public Power Supply System	1984
Richland	WPPSS No. 4	1,250,000	Washington Public Power Supply System	1984
Satsop	WPPSS No. 5	1,242,000	Washington Public Power Supply System	1985
Sedro Woolley	Skagit Nuclear Project: Unit 1	1,277,000	Puget Sound Power & Light	1985
Sedro Woolley	Skagit Nuclear Project: Unit 2	1,277,000	Puget Sound Power & Light	1987

WISCONSIN

La Crosse	La Crosse (Genoa) Nuclear Generating Station	50,000	Dairyland Power Cooperative	1969
Two Creeks	Point Beach Nuclear Plant: Unit 1	497,000	Wisconsin Michigan Power Co.	1970
Two Creeks	Point Beach Nuclear Plant: Unit 2	497,000	Wisconsin Michigan Power Co.	1973
Carlton	Kewaunee Nuclear Power Plant: Unit 1	535,000	Wisconsin Public Service Corp.	1974
Site not selected	Haven Nuclear Plant: Unit 1	900,000	Wisconsin Electric Power Co.	1987
Site not selected	Haven Nuclear Plant: Unit 2	900,000	Wisconsin Electric Power Co.	1989
Durand	Tyrone Energy Park: Unit 1	1,150,000	Northern States Power Co.	1985

PUERTO RICO

Arecibo	North Coast Power Plant	583,000	Puerto Rico Water Resources Authority	Indef.

Theft & Sabotage Threat

Security, Safety & Transportation Problems

With the increased use of nuclear power and the spread of terrorism, many observers warned of the possibility of sabotage and of the theft of radioactive material for the illicit manufacture of atomic bombs. By late 1970 tighter security rules and safety standards were imposed to regulate the guardianship and transportation of radioactive material. The new regulations were issued following reports of plutonium disappearing and continued charges that security measures at nuclear plants were inadequate.

Nuclear theft threat seen as great. A government report prepared by an AEC official and several outside consultants was released April 26, 1974, The report stated that the danger to the public posed by theft of nuclear materials was greater than that from any "plausible" power plant accident. Criticizing the inadequacy of security measures used by private industry, the report recommended creation of a federal force to guard and transport uranium and plutonium.

The report was released by Sen. Abraham Ribicoff (D, Conn.) despite a plea by the AEC that it not be made public, it was reported April 27.

Another study, sponsored by the Ford Foundation, had reached similar conclusions and noted that an illicit atomic device could be "relatively easy to make." The study, reported April 7, also recommended creation of a federal security force.

The New York Times reported May 2 that the AEC had begun ordering utilities to hire armed guards for their plants, despite objections from some companies that security should be a government function, and that the weapons used by private guards might pose new dangers. According to an AEC official, the commission had decided that if nuclear power were to be developed privately, the industry should be responsible for security.

A-bomb guards told: 'shoot to kill'. Personnel guarding U.S. atomic weapons were ordered to "shoot to kill" to protect the weapons, a Pentagon spokesman had acknowledged Oct. 11, 1974.

The orders were a direct result of American troops being put on alert during a crisis in the Middle East. The AEC and the Pentagon issued the orders as a precaution against theft by Arab terrorists.

"They [the guards] should be informed that the loss of a nuclear weapon or device through theft or sabotage would have the most serious consequences. . . . They must understand that to prevent the illicit loss or destruction of nuclear weapons they are expected to discharge their firearms with the intent of hitting and if necessary killing the person or persons being fired upon. . . ," the order read.

A-cargo transport curbed. The AEC and the Department of Transportation (DOT) had issued new rules April 2, 1973 to regulate ground and air transport of radioactive materials.

The ruling would transfer to the AEC control of handling and transport of all fissionable material and large packages of other materials. DOT would retain control of small packages, for which it would design new handling safety standards.

The AEC had announced a ban Jan. 30 on air transport of significant amounts of fissionable uranium and plutonium.

AEC plans tighter safety rules. A plan submitted by the AEC Nov. 7, 1974 provided for greater security for nuclear power plants and fuel transporters to prevent the sabotage and theft of nuclear materials.

The proposed measures, tightening regulations that had gone into effect in December 1973, were made effective after appeal periods of 30–60 days. They required each of the 50 commercial nuclear power plants in the U.S. to have a security organization that would include monitored isolation zones around fences and bullet-proof windows and doors for reactor control rooms. Specially designed trucks and armored cars would have to be used by firms shipping large amounts of uranium and plutonium; each vehicle was to carry two armed guards, accompanied by one or two escort vehicles carrying armed guards.

FBI warns of atomic thefts. Officials of the Federal Bureau of Investigation (FBI) warned Jan. 3, 1975 of the possible theft of atomic material from nuclear plants by terrorists for use in the manufacture of crude atomic bombs. In the past year the FBI had investigated seven letters threatening to explode nuclear bombs in Boston, Des Moines, San Francisco and Lincoln, Neb., according to agency officials.

The FBI warning followed reports by the AEC and private nuclear plants that thousands of pounds of enriched uranium and plutonium usable in the making of nuclear bombs were unaccounted for, according to a New York Times article published Dec. 29, 1974.

A-fuel reported missing. A report by the General Accounting Office (GAO) cited Aug. 5, 1976 at a session of the House Small Business Committee's subcommittee on energy and the environment said that government nuclear facilities were unable to account for several tons of nuclear fuel that could be used for making weapons.

The U.S. government had maintained that most of the missing fuel could be accounted for by material embedded in reprocessing machinery and by crude statistical controls. However, the classified GAO report, which was disclosed during testimony, charged that lax government inventory controls and security measures were responsible for the loss of as much as 6,000 pounds of weapons-grade plutonium and uranium. It had been reported earlier that there were two known instances where government employes had smuggled out of guarded facilities enough special nuclear materials to develop nuclear weapons.

The U.S. disclosed later (Aug. 4, 1977) that 8,000 pounds of material used to make atomic bombs—4,800 pounds of enriched uranium and 3,400 pounds of plutonium—were missing or unaccounted for as of September 1976.

Officials of the NRC and the ERDA had minimized the possibility that the bomb-grade material had been stolen. The officials said that losses of uranium and plutonium in machinery and wiping cloths and as scrap, coupled with sloppy accounting, probably explained the missing material.

The announcement marked the first time the government had made public detailed information on losses of fissionable material.

Several members of Congress challenged the assessment of the problem by NRC and ERDA, saying they would investigate whether unauthorized parties had acquired any of the missing material. Rep. John Dingell, (D, Mich.) said an audit by the General Accounting Office showed the losses to be higher than NRC and ERDA contended.

Dingell Aug. 8 chaired a hearing of the House Interstate and Foreign Commerce subcommittee on energy and power at which ERDA officials admitted there were another 16 tons of enriched uranium that could not be accounted for. The material was from the facilities at Oak Ridge, Tenn. and Portsmouth, Ohio, where atomic weapons were manufactured.

The 16-ton figure had not been included in the Aug. 4 report, ERDA Administrator Robert Fri said, because it was only a rough estimate. Fri said the weapons-grade material was not actually lost or stolen; instead, it was just the uranium that had condensed along the hundreds of miles of pipes at each facility. (The uranium was diffused through the pipes in gaseous form in order to separate various isotopes.)

Fri said about another 64 tons of non-weapons-grade uranium was believed to have accumulated in the pipes of the two facilities.

Fri also testified that investigators of every suspicious incident involving missing weapons-grade material had always concluded "there is no evidence that a significant amount of special nuclear material had been stolen or diverted."

After the hearing, however, subcommittee staff director Michael Ward told reporters that officials in the U.S. intelligence community had "strong suspicions that a diversion occurred."

NRC reviews plutonium transit. A June 2, 1975 Wall Street Journal report said that the NRC was reviewing the regulation of air transport of plutonium and enriched uranium to see if safety precautions were adequate.

The NRC said that about 2.2 million packages of radioactive materials are shipped in the U.S. annually.

A disclosure in March that two shipments of plutonium, 100 pounds each, had been flown into New York City's John F. Kennedy Airport evoked protest in the city, state and Congress. The shipments were transshipped from Kennedy by truck to Pennsylvania. A third scheduled shipment, of 290 pounds of plutonium oxide, routed via Kennedy to Rochester, N.Y., was canceled by the ordering company April 4 as a result of the outcry.

The NRC later disclosed in an "Annual Report" to Congress (1977) that pending a joint study by the commission and the Chicago mayor's office, shipments of highly enriched uranium would not be shipped through Chicago's O'Hare Airport, but would be moved through other airports as of December 1977.

As a result of the outcry, the New York City Board of Health Jan.15, 1976 passed an amendment to the Radiation Control section of the New York City Health Code regulations. The amendment severely restricted the shipments of certain radioactive materials, including plutonium, into the city.

To investigate the regulatory aspects of the New York City radiation control laws, Senate hearings were held in 1978 by the Committee on Commerce, Science & Transportation.

Dr. Leonard R. Solon, Director of the Bureau for Radiation Control of the City's Department of Health, testified Aug. 16, 1978 before the committee's Subcommittee on Science, Technology & Space and it's Subcommittee on Surface Transportation. Solon defended the city's Health Code regulations, which, he said, "turns out to be a precise regulatory mechanism . . . protecting the health and safety of New York City residents by controlling dangerous shipments of . . . [radioactive material]." Solon added that the "dispersion of even a small fraction of . . . one of these shipments as the result of an air crash . . . could have cataclysmic results bringing death or serious injury to thousands of New Yorkers. The inhalation or ingestion of milligram quantities of plutonium would result in almost certain death . . . within weeks or months. Inhalation of a quantity one hundred fold smaller—tens of micrograms—would result in . . . a susceptibility to fatal pulmonary cancer in one or two decades. To recapitulate, one milligram, death—10 micrograms, lung cancer—one air shipment, 40 million milligrams or 40 billion micrograms."

A-material transit accident. A cargo plane enroute from New York to Washington crashed near Washington National Airport July 2, 1970 killing the two-man crew. Among the twin-engine plane's cargo were two 35-pound canisters containing radioactive isotopes used for medical diagnosis by Walter Reed Army Medical Center and the National Institutes of Health.

A-industry charged with security laxity. James Conran, a member of the NRC staff, criticized the commission's safeguards programs and policies. Conran also charged that because of his disagreement with NRC policies, he had been involuntarily transferred to a new position.

In response to these allegations, an investigative hearing was held July 29, 1977 by the Subcommittee on Energy & the Environment of the House Committee on Interior & Insular Affairs.

An NRC task force had been formed in April to review Conran's charges. The task force report supported some of the allegations, disagreed with others and made several recommendations that the commission later adopted. In a summary statement submitted to the hearing, the task force noted these allegations made by Conran: (a) existing safeguards at NRC-licensed facilities are afflicted pervasively by serious and chronic weaknesses that pose potential hazards to public health and safety; (b) there is a general limitation, perhaps suppression, of information, especially that bearing on the questions of the relative ease and the likelihood of success in design and fabrication of a clandestine fission explosive (CFE Information); (c) ERDA and NRC officials made misleading and untrue statements concerning the specific categories of information; (d) the current safeguard regulatory practice includes a willingness to compromise safeguard effectiveness in favor of costs, image and acceptance by the industry, and a failure to account properly for uncertainties, such as "the design threat."

After reviewing the charges, the task force said it had found "limitations on the availability of information both to and within NRC. However, we have found no case in which such limitations resulted in a . . . compromise of safeguards sufficiency. Some limitations were the result of protecting classified information. Other limitations appear to have been the product of the natural and expected difficulties of a new agency in its formative stages, and the task force has concluded that for the most part appropriate corrective steps have been, or are being, initiated."

Among the recommendations made by the task force were that the NRC: (a) "ensure that all statements on missing or stolen nuclear materials reflect the uncertainties of material control and accounting technology"; (b) "promptly initiate a deliberate and detailed review of the onsite CFE scenario described by Mr. Conran to determine its relevance to the design and operation of safeguards systems"; (c) "expeditiously establish improved guidance for the acceptable level of protection against theft."

A-worker firing to be probed. The NRC said March 29, 1978 it would look into the dismissal of a construction worker at an atomic plant in Missouri to determine if he had been fired in retaliation for bringing certain safety questions to the attention of federal officials.

The worker—William Smart—had charged that the construction firm he had been employed by, Daniel International, had used bad building techniques, had violated safety rules and had allowed defective parts to be installed. Daniel International was one of the companies building the Callaway nuclear power plant at Fulton, Mo.

NRC officials said that Smart's safety criticisms had been found to be virtually groundless. The NRC was looking into his dismissal, however, to make sure that workers at atomic facilities were not punished for drawing attention to safety problems, the officials said.

Daniel International said Smart was fired for refusing to perform work he had been ordered to do. Smart claimed the dismissal was in retaliation for going to the NRC.

Rep. William Clay (D, Mo.) said in a statement that Daniel International's account of the dismissal "rings hollow," the UPI reported March 29. Clay said Smart was fired for "whistle-blowing about faulty construction and quality control."

Smart and local environmentalists were seeking a review by the General Accounting Office of Smart's charges. Clay said he also favored a GAO review.

International shipments & safeguards. The NRC was reported to have suspended issuance of import and export licenses for nuclear materials March 28, 1975 pending a policy review because of growing concern about such shipments.

The European Community protested April 11 that the decision, made without prior consultation with it, could seriously affect operations of the community's nuclear power stations.

NRC Commissioner E. A. Mason made a formal statement to a European nuclear conference in Paris April 22 that no U.S. decision had been made to delay or suspend shipment of enriched uranium. He said the NRC commissioners themselves were required by regulation to process the licensing for handling a shipment of nuclear supplies, a procedure previously handled by its predecessor, the AEC, on the staff level. This had been "erroneously reported" as a suspension of licensing, he said.

The same day, the NRC announced approval of the export of 1.1 million pounds of uranium ore for use in West Germany nuclear power reactors.

The Paris conference, the first European conference on the peaceful uses of atomic energy, organized by the newly founded European Nuclear Society, was attended by 3,000 delegates from 47 countries. It ended April 25 with an expression of approval from Andre Giraud, head of France's Atomic Energy Commission, for consideration in the future of the "social costs" of development of nuclear

power programs. He supported a suggestion that various alternative "scenarios" of development be proposed to permit wider public participation in the decisions made, such as for siting of power stations.

A group of experts, convened by the International Atomic Energy Agency from 11 countries, met in Vienna May 2 and proposed stricter international procedures for protecting nuclear material. It endorsed in principle, without details, use of armed guards to protect nuclear materials from theft or sabotage. It also advocated consideration of an international convention to set standards for guarding international transportation of the materials, and it recommended protection for shipments in amounts as small as 10 grams of plutonium. Under the old standards, protection was recommended for enriched uranium in amounts of one to five kilograms.

A 60-nation review conference of the treaty for preventing the spread of nuclear weapons issued in Geneva May 30 a caution to countries with nuclear installations to take steps for stricter physical security to prevent theft of fissile materials.

Plutonium air transit ban. President Gerald R. Ford Aug. 9, 1975 signed into law a bill authorizing $222.9 million in funds for the NRC for fiscal 1976. The authorization bill, approved by both houses of Congress by voice vote July 31, contained a provision banning the air transport of plutonium until a rupture-proof container could be developed to make the material secure even in the event of an airplane crash.

A-plants told to tighten security. The NRC Feb. 22, 1977 ordered the 63 federally licensed nuclear power plants to take major new safety precautions against terrorist attacks.

The order required each facility to hire at least 10 guards or trained personnel armed with semiautomatic rifles. It required guards to respond to all threats with an equal degree of force, including shooting to kill if necessary. Most plants currently were protected by three or four guards with pistols.

The order also required implementation of elaborate procedures to limit employe access to certain vital areas of the plants and construction of a bullet-resistant alarm center to ensure that a call could be made for outside assistance in case of attack.

Most of the new measures were to be taken within 90 days, but any construction required would not have to be completed for 18 months. Plants not in compliance by the deadlines would be closed down. An NRC spokesman estimated the cost to each facility at about $1.5 million to $2 million a year, compared with the $500,-000–$750,000 most plants currently spent on safety.

The new regulations were designed primarily to protect against sabotage by a group of "several" outsiders cooperating with at least one employe of a plant. NRC officials said they did not know of any group in the country with the motivation and skill to attack a nuclear reactor and cause an accident that would release deadly radioactivity into the air. However, they said the number of threats against nuclear facilities had increased to 68 in 1976 from an average of about 16 a year through 1975.

The threat of thefts of plutonium, the raw material of atomic bombs, was considered remote at nuclear reactors, officials said, because the plutonium was contained only in burned fuel rods at the core of the reactor and was too hot to touch. Nevertheless, the NRC was considering other safety measures for the 13 facilities licensed to process the plutonium and enriched uranium that was used mostly for to fuel reactors powering Navy ships.

The NRC had been petitioned by the Natural Resources Defense Council Feb. 2, 1976 to initiate a number of emergency steps to tighten the existing safeguards at the 16 research and processing plants licensed to use weapons-grade nuclear materials.

The council, a private environmental organization, considered current security inadequate. Among the council's proposals were use of federal marshals at the plants and possible repossession by the federal government of all weapons-grade nuclear materials.

'Violence' in anti-nuclear movement. Rep. Larry McDonald (D, Ga.) told the House of Representatives Nov. 3, 1977 that "the role of violence in the antinuclear power . . . [movement] has become an important topic of debate within activists circles . . . in the U.S. . . . Western Europe [and Latin America]. The line of argument typified by Clamshell Alliance spokesman Sam Lovejoy is that 'nonviolence' includes sabotage and other forms of property destruction, and that 'violence' only includes deliberate attacks on humans and animals."

McDonald continued: "A wide variety of revolutionary groups are seeking to penetrate the antinuclear power movement both in the United States and in Europe." . . . The mass demonstrations in Europe have been the occasion for contingents of revolutionaries to battle police . . . The issue of terrorist violence against nuclear power plants has been joined in this country by the New World Liberation Front (NWLF) the reincarnation of the Symbionese Liberation Army which has been active in California, Colorado, and Oregon. . . ."

McDonald cited the terrorist bombing Oct. 10, 1977 of the Trojan nuclear power plant visitor's center in Ranier, Ore. The bombing was carried out by the Environmental assault Unit, a section of the NWLF and was directed against the construction and development of nuclear power in the U.S.

McDonald quoted The Information Digest as reporting . . . Samuel Holden Lovejoy [of the NWLF] . . . injected the question of violent 'direct action' against nuclear power plants into the anti-nuclear environmentalist movement when in February 1974 he released the guy ropes causing the collapse of a $50,000 preliminary weather tower at a nuclear plant construction site near Montague, Ma." This was the first reported act of terrorism against an operating U.S. nuclear-power facility, but several such attacks had preceded it in Western Europe and Latin America.

Among other terrorist attacks: (a) The Argentine Atucha power plant was invaded in March 1973 by 15 members of the People's Revolutionary Army (ERP) a Trotskyist terrorist group. (b) A bomb was detonated in May 1975 at the Fessenheim plant in West Germany by the Red Army Fraction (Baader-Meinhof group). (c) Three months later, two bombs were set off in Brittany, France at the Monts Aree plant site by the Breton Liberation Front.

Terrorists attack French power plants. Several bomb and machine-gun attacks took place in France Nov. 19, 1977. Most were directed at the state electricity company and firms working for the Defense Ministry.

A policeman was seriously injured at La Cappelle Marival in central France when three of the bombs exploded at the police station. The atomic physics laboratory at Toulouse University also was severely damaged in the spate of bombings.

An organization called the "Carlos Committee" claimed responsibility for the attacks Nov. 21. The group reportedly was an independent terrorist organization opposed to the development of nuclear power in France.

In another bomb attack Dec. 19, two explosions wrecked Fauchon, the famous Parisian food shop. There were no deaths or injuries, but the building housing the store was burned out and nearby structures were slightly damaged.

The store had received no threats or demands before the bombing. Extreme leftists often referred to Fauchon as an example of "bourgeois decadence." A Maoist group had raided the store in 1970 and handed out several truckloads of its food in the slums.

A-sub theft foiled. A plot to seize a nuclear submarine from the New London, Conn. naval yard resulted in the arrest Oct. 4, 1978 of Edward J. Mendenhall, Kurtis J. Schmidt and James W. Cosgrove.

The three men had schemed to recruit and train a 12-member crew to board a submarine tender alongside the U.S.S. *Trepang,* sink the tender with explosives, and, in the subsequent confusion, kill the estimated 107 members of the submarine's crew.

The conspirators then planned to move the submarine out of the harbor. As a diversion, they had considered firing a missile at New London "or one of several principal East Coast cities," the Federal Bureau of Investigation said. The *Trepang* carried an anti-submarine missile with a nuclear warhead having a range of about 30 miles (48 kilometers). The scheme then called for sailing the submarine out into the Atlantic Ocean where it would have been turned over to a purchaser.

The FBI, alerted July 26 by a person the three had attempted to recruit, said there was no indication that there was a foreign power or any kind of subversive group involved in the plot. The FBI made the arrests when it was decided the plot had gone far enough to constitute a conspiracy to steal.

The conspirators had told an undercover FBI agent that they had the talents available and planned to ask $200 million for the submarine. The FBI confirmed that Cosgrove had served in the navy and attended submarine school in New London. An anonymous source, quoted in the St. Louis Globe-Democrat, said Cosgrove had served aboard the submarine from September 1973 to November 1974.

However, a navy spokesman said Oct. 5 that it would take a highly skilled crew of at least 33 men to get the submarine under way. Earlier it had been estimated that it would be impossible without a trained crew of 100. Navy officials said they were so sure of security measures at the New London base that no extra security was called for even after the FBI had alerted them to the plot.

U.S. Rep. Christopher Dodd (D, Conn.) said Oct. 5 after a briefing by Navy and FBI officials that he was assured the base was secure. "It's not like you have the keys, get in and drive down the Thames River," he said.

High seas uranium loss confirmed. The European Community (EC) Commission May 2, 1977 confirmed reports that 200 metric tons of uranium ore had mysteriously disappeared from a freighter bound from Antwerp, Belgium to Genoa, Italy in November 1968. The incident had been reported April 28 by the Los Angeles Times and the New York Times and on April 29 by Paul Leventhal, a former

Senate Government Operations Committee staff expert on the spread of nuclear weapons.

Leventhal, in a speech to the Conference for a Non-Nuclear Future in Salzburg, Austria, said that several weeks after the ship had failed to make its scheduled arrival in Genoa, it had "reappeared with a new name, new registry, new crew, but no uranium." (The conference was timed to coincide with a two-week conference on nuclear power, also in Salzburg, sponsored by the International Atomic Energy Agency.)

Though investigations by at least four nations had never officially resolved the mystery of the disappearance, some U.S. and European intelligence officials reportedly were convinced that the uranium had found its way to Israel. EC officials privately indicated May 2 that this was their understanding as well.

The Israeli Atomic Energy Commission April 29 denied that Israel had any connection with the disappearance of the uranium.

The reports of the missing uranium again raised speculation as to whether Israel had the capability to produce nuclear weapons and, if so, whether it had already done so. The Israeli Atomic Energy Commission spokesman, in making the denial April 29, refused to say where Israel had obtained the uranium for its top-secret nuclear reactor at Dimona in the Negev Desert. Built in the 1950s with French assistance, the Dimona reactor had never been opened to inspection by foreigners.

EC officials also privately alleged that the leak of the nine-year-old story in the U.S. press had been a deliberate tactic by U.S. officials to give substance to their current campaign to get Europeans to accept stricter international controls on the sale and transport of nuclear materials. One EC official said that if the story had indeed been leaked, "it was a dirty blow."

Atomic Accidents & Liability

The Liability Issue

Few penalties for violations reported. Reporting on a study of AEC records, the New York Times said Aug. 25, 1974 that the commission had imposed penalties on only a small fraction of nuclear installations at which violations had been found, despite the fact that many of the violations could have created significant radiation hazards.

According to records for the year ended June 30, the AEC found 3,333 violations in 1,288 of the 3,047 installations inspected. The commission imposed punishment in eight cases: license revocations involving two companies and civil penalties totaling $37,000 against six others.

Under the AEC's three-level classification of the seriousness of violations, 98 were in the top category (violations which had caused or were likely to cause radiation exposures in excess of permissible limits). In category two—defined as violations that, if not corrected, might lead to exposures above permissible limits—there were 2,132 violations. Category three—infractions involving documentation and procedural matters—included 1,103 violations.

The report also noted that a composite ratio of violations and penalties over the past five years was similar to that of the most recent fiscal year: 10,320 inspections; 3,704 installations with one or more violations; and 22 cases involving imposition of penalties.

AEC cites hazards at Oklahoma plant. A report released Jan. 7, 1975 by the AEC upheld several charges made by the Oil, Chemical and Atomic Workers Union of health hazards and other dangers at the Kerr-McGee Corp. in Crescent, Okla.

The report, based on the findings of AEC investigators, substantiated 20 of the 39 union claims of danger to the health of the firm's workers and said some X-ray negatives of fuel rods being manufactured for a plutonium-powered reactor had been falsified and that some data concerning the rods had not been used properly.

As a result of the inquiry, the AEC criticized Kerr-McGee for neglecting to report to the commission a plutonium processing equipment failure that had closed the plant for 40 hours, permitting on two occasions an excess amount of plutonium in a specific work area and using a small amount of plutonium in an unauthorized form.

Among the health hazards cited were the sending of a worker into a dangerous area without informing him that a respirator was required to protect him from possible radiation, failure to make certain that the respirators were working and "errors" that resulted in contamination. Since Kerr-McGee started its plutonium operations in 1970, 17 safety lapses in which 73 employes had been contaminated were reported, according to AEC records.

A previous AEC report released Jan. 6 said an employe and critic of Kerr-McGee's safety measures, Karen G. Silkwood, had swallowed microscopic amounts of plutonium seven days before she had been killed in a car accident Nov. 13, 1974. According to the AEC's findings, Miss Silkwood's contamination "probably did not result from an accident or incident within the plant"; two urine samples submitted by Miss Silkwood "contained plutonium which was not present when the urine was excreted" but that the subsequent commission inquiry "did not establish how or by whom the radioactive material was added."

Plant critic killed—A critic of the safety procedures at the Crescent, Okla. plant was killed in an automobile crash Nov. 13. The death of Karen Silkwood, 28, a technician at the plant, and several accidents at the plant were being investigated by the federal government.

Miss Silkwood's car crashed into a culvert as she was driving to a meeting with an official of the Oil, Chemical and Atomic Workers Union and a New York Times reporter with information of the alleged failure of the company to protect workers from the lethal dangers of plutonium radiation. She and two co-workers had criticized the firm's safety practices at a meeting with the AEC in Washington Sept. 27. Silkwood had been exposed to a large amount of radiation in a plant accident a week before her death.

Officials of the Oklahoma State medical examiner and the state highway patrol said Miss Silkwood's death was accidental and denied the union's contention that her car crashed into the culvert after being struck from the rear by another car.

The Federal Bureau of Investigation launched an investigation into the accident after the union asked Attorney General William B. Saxbe Nov. 18 to order an inquiry. A private investigator hired by the union submitted the results of his study Dec. 23, which upheld the union's original contention that Miss Silkwood's car had been struck from the rear and disputed the state's view that no other vehicle was involved in the mishap.

The AEC started an investigation into an allegation that the plutonium factory had been falsifying the inspection of fuel rods to be used in an experimental reactor and into the plutonium accident involving Miss Silkwood. Three subsequent incidents at the plant Dec. 16 prompted the commission to widen its investigation and forced Kerr-McGee to close the facilities "until corrective action has been taken."

Court rulings on nuclear liability limit. The Supreme Court ruled unanimously June 26, 1978 that the Price-Anderson Act was constitutional. The law set a ceiling of $560 million on liability for all deaths, injuries and property damage arising from a single accident at a private nuclear power plant. The statute was opposed by environmental groups that charged that it violated the due-process guarantee of the Fifth Amendment.

The case was *Duke Power Co. v. Carolina Environmental Study Group.*

The Supreme court decision overturned a lower court ruling Mar. 31, 1977 by Judge James McMillan which had declared the Price-Anderson Act unconstitutional.

The Price-Anderson Act, passed in 1957 and extended in 1975 until Aug. 1, 1987, had set up a joint private-federal insurance program for the nuclear power industry. It was challenged in 1973 by Ralph Nader's Public Citizens Litigation Group on behalf of the Carolina Environmental Study Group and 40 individuals living near the Duke Power Co.'s McGuire Nuclear Station on Lake Norman in North Carolina. The defendants in the case were Duke Power and the U.S. Nuclear Regulatory Commission (then the Atomic Energy Commission).

Judge McMillan ruled that the act violated constitutional guarantees of equal protection under the laws "because it provides for what Congress deemed to be a benefit to the whole society (the encouragement of the generation of nuclear power) but places the cost of that benefit on an arbitrarily chosen segment of society, those injured by nuclear catastrophe."

William Schultz, attorney for the Nader group, said the nuclear industry had wanted the Price-Anderson Act as protection for its investments. He said that if the judge's ruling stood, the effect could be to slow nuclear power development because it was unlikely that utilities would want to assume unlimited liability. (Rising construction costs and the unexpectedly slow rise in demand for electricity already had caused cancellation of several nuclear power plant projects.

Price-Anderson Act amendment proposed—In view of the high court decision upholding the Price-Anderson Act, Rep. Ted Weiss (D, N.Y.) asked Congress July 13, 1978 to remove the act's liability limit. Weiss said that "prompt and affirmative congressional action" was needed "to protect fully all potential victims of a nuclear disaster."

Weiss introduced an amendment that would "require all atomic plant licensees to arrange a combination of individual private policies and an industry-financed insurance pool to cover the balance of claims. . . ." Weiss said his amendment to the Price-Anderson Act:

"Stipulates that each licensee must obtain the maximum coverage available from private insurers;

"Provides that in the event of an accident at a particular plant, this facility's insurnace would be the first source of claim payments;

"States that if the judgment against a given facility exceeds its private insurance coverage, the plant would then be liable up to the total of its assets;

"Requires that if a court determines that the plant's assets are insufficient to meet the full judgment, the industry pool shall cover the balance of outstanding claims;

"Mandates that pool participants shall contribute to the cost of the judgment according to a formula based on each licensee's generating capacity, its degree

of risk, its total assets and other factors; and

"Provides that should a court determine that a given plant cannot meet its proportional share to the pool, that plant would then become eligible for long-term Federal loans to be secured against the plant's assets;

"Makes explicit that each licensee shall be liable for any incident at its facility resulting in injury to persons and/or damage to property; and

"Waives the 20-year statutory limitation on claims resulting from a nuclear incident. . . ."

Weiss continued:

"Unlike the Price-Anderson Act which mandates an automatic Federal subsidy for nuclear power insurance, my bill would entail use of Federal funds only in the event of a major catastrophe where victims' claims exceed a plant's private insurance coverage, a plant's total assets and the proportional contribution toward an industrywide pool. Even then, Federal moneys would be used only to provide interest-bearing loans.

"If nuclear energy is indeed as safe as its proponents maintain, then the nuclear industry should be able to convince private insurers to write policies in excess of the current $140 million average amount. Atomic energy advocates should also be willing to back up their claims of a nearly infallible energy source by financing a pool to cover the costs from an accident whch they regularly describe as 'almost impossible.'

"Price-Anderson was enacted as a means of encouraging private industry to accelerate nuclear power development. That was 21 years ago, and since then more than 60 atomic powerplants have been activated in our Nation. The time has come for nuclear power either to stand or fall on its own merits. No other method of energy-generation enjoys a similar liability limitation, and Congress now has a clear duty to end this unfair underwritng of atomic power. . . ."

Insurers request safety evaluations of 25 plants. Ralph Nader released Aug. 13, 1975 confidential documents disclosing that a pool of insurance companies had requested safety evaluations at 25 nuclear plants. The companies were members of the Nuclear Energy Liability and Property Insurance Association (Nelpia), which issued liability insurance against nuclear hazard to the 55 nuclear plants in the U.S. It held property insurance coverage for 25 plants, those for which it requested the safety reports.

The request followed a fire in March at the Browns Ferry nuclear power plant in Alabama. A Nelpia report estimated the damage from the fire to be at least $50 million, which made it "the most expensive fire in history," Nader said. The plant was expected to be shut down for repairs through December.

A petition from 31 consumer and environmental groups was sent Aug. 7 to the NRC requesting it require public evacuation drills to prepare for possible nuclear power plant disasters. The petition was announced by Nader.

Accidents

3 die in military reactor blast. Three military technicians were killed when a nuclear reactor exploded at the government owned (ARCO) National Reactor Testing Station in Idaho Falls, Ida. Jan. 3, 1961. John A. McCone, AEC commissioner, said that the resultant escape of radioactivity was "largely confined in the reactor building." The New York Times reported Jan. 20 that the blast was caused by a nuclear "runaway"—escaping radioactive neutrons that had come in contact with gold and copper objects, converting the metals into radioactive forms. The testing station had been in operation for 11 years. The deaths were the first reported from a reactor accident.

Rocky Flats A-plant fire. The AEC denied Feb. 19, 1970 that the plutonium released during a fire in 1969 at its Rocky Flats atomic warhead plant near Golden, Colorado constituted a health hazard. The commission admitted however, that minute traces of the plutonium had leaked from waste storage barrels. It said a new waste treatment plant would be installed.

The statement was in answer to a report

sent to AEC Chairman Glen T. Seaborg Jan. 13 by the Colorado Committee for Environmental Information. The report had said that soil samples taken near the site, 16 miles from Denver, indicated an amount of plutonium that could be stirred up by a strong wind and carried off to be inhaled by people in the area.

In response to the AEC's contention that the quantities of plutonium were "miniscule," Dr. E. A. Martell, a member of the information committee, said that "each plutonium particle in the lung produces millions of times more radiation to the tissue around it than a dust particle carrying natural radioactivity."

The information committee had disputed the AEC's contention that no radioactivity was vented during the fire.

The EPA reported Dec. 5, 1974 that cattle grazing on land east of the Rocky Flats atomic weapons plant showed high degrees of plutonium in their lungs.

The AEC had said May 20, 1969 that the Rocky Flats fire had been caused by the spontaneous ignition of plutonium located in a production building. It also reported that the fire had not caused any injuries and that radioactive contamination had been confined to the vicinity of the building. Damages had been estimated at $40–$50 million, but the cost of the plutonium was not included in the estimate.

The AEC later approved a $130 million renovation of the Rocky Flats plant, it was reported Jan. 4, 1974. The improvements were designed to limit the risk of fire or radiation leaks.

Rocky Flats plutonium debate continues. Criticizing the plutonium health hazards associated with the Rocky Flats atomic weapons plant, Rep. Richard L. Ottinger (D, N.Y.) June 16, 1978 inserted in the Congressional Record a *Saturday Review* article written by Peter J. Ognibene on the issue.

Ognibene said soil samples taken near the Rocky Flats plant site were found by Dr. Carl J. Johnson, a public health official, to be "as high as 3,390 times the [normal] background radiation" level. Johnson's samples were taken near the surface—no deeper than half a centimeter. Rocky Flats scientists however, take soil samples two inches deep. According to Johnson, this "dilute[s] the respirable plutonium particles with dirt."

Controversial studies discuss the possible genetic effects of plutonium contamination. Dr. John Cobb, a professor at the University of Colorado Medical Center, asserted that "plutonium goes to the testes and ovaries and probably damages germ cells, but we don't know. . . ." Health researchers at the University of Denver found that "workers at Rocky Flats who had low levels of internal plutonium exhibited certain effects. 'We were somewhat surprised,' the researchers reported, 'at the elevated prevalence of chromosome aberrations in the one to 10 percent . . . (body burden) individuals.' Federal regulations establish a limit on the amount of plutonium a worker may have in his body; it is called the maximum permissible body burden." The researchers "also found 'a marked increase in the prevalence of total [chromosome] structure aberrations in the (greater than) 50 percent . . . [maximum permissible body burden] group of men.'" In the absence of medical "studies on the offspring of nuclear workers" and adequate health records, however, no one can "determine the genetic consequences to children whose parents were occupationally exposed to radiation," the Denver researchers said.

Plutonium falls into ocean from spacecraft. During an aborted spaceflight April 17, 1970, the lunar module from the Apollo 13 was jettisoned and splashed down into deep water in the Pacific Ocean, northeast of New Zealand. A Snap-27 nuclear generator loaded with 8.6 pounds of radioactive plutonium was reported to have fallen from the lunar module.

Test accident disclosed. The AEC said Nov. 1, 1973 that an accident had occurred Oct. 20 at its experimental gas-centrifuge uranium production facility at Oak Ridge, Tenn. The device being tested was part of a highly-secret method of separating uranium 235 from other uranium products more efficiently than by other methods.

An AEC spokesman said parts of one centrifuge, spinning at high speed, had broken into pieces and damaged nearby centrifuges. The spokesman said it would be inaccurate to call the accident an explosion. Other less serious accidents had occurred before, he said, but had not been made public.

Browns Ferry fire probe. The official investigation of a March 22, 1975 fire at the world's largest nuclear power plant at Browns Ferry, Ala. found Feb. 28, 1976 that federal attention to fire prevention and control was lacking.

The report, prepared by a committee of the NRC, concluded that there were "lapses in quality assurance in design, construction and operation" of the two reactors at Browns Ferry, which were operated by the Tennessee Valley Authority.

TVA was faulted on its design of the plant and the way it managed the fight against the fire; the NRC itself was blamed for failing to properly exercise its responsibility "to assure that the licensee operates the plant safely."

The fire ignited from a small candle flame held by an electrician checking for air leaks in electrical cable. The cable's insulation caught fire and the fire spread to the electrical control room and through it to the room housing the nuclear reactor. There, the fire was put out with water, but only after an effort to contain it with dry chemicals, an incorrect procedure said to have prolonged the fire.

One of the two reactors lost some of its nuclear cooling water and an emergency cooling system was disabled. Other pumps were operated to stabilize the reactor core situation. The fire did not damage the reactor cores, and no radiation leaks were reported.

Smoke detectors in the control room failed to work, the report said, and the design of the detection and control system was sorely deficient. No detectors were installed in the area adjacent to the control room, through which the fire spread, and wires for both the main and back-up detection systems were placed alongside one another. They burned together. Other wires improperly close together short-circuited from the fire's heat and prevented operation of valves and pumps to protect the nuclear cores.

The Browns Ferry nuclear power plant had been closed since the fire. It was expected to cost TVA customers $150 million by the time the reactors were back in operation.

Following the Browns Ferry plant fire, the NRC reviewed the fire protection programs of all operating plants, sending teams of specialists to each facility. One of the results was that improved guidelines were issued by the commission for fire protection in both new and existing nuclear power plants in the U.S.

Scientists report on Browns Ferry Fire— The Subcommittee on Energy & the Environment of the House Interior & Insular Affairs Committee June 11, 1976 held an oversight hearing on nuclear reactor safety. At this hearing, the Union of Concerned Scientists submitted its independent probe of the Browns Ferry fire and its safety systems failure.

The scientists' report said that the accident "revealed a surprising and ominous frailty in one of the world's largest nuclear plants: elaborate safety systems installed to control major accidents can be rendered inoperable by the very accident they are intended to control. . . ."

The report continued:

"The Browns Ferry fire created a type of nuclear plant accident known as common-mode failure. A common-mode failure involves a single event that causes multiple failures of plant components or systems. Since nuclear power plants . . . rely heavily on . . . redundant safety equipment, . . . common-mode failures can seriously jeopardize plant safety by creating an across-the-board safety system failure. . . .

"According to the NRC's investigation, [the accident] was the direct result of serious safety deficiencies at the plant . . . [that included]":

(a) The presence of highly combustible polyurethane foam in the plant's electrical control system; (b) lax management controls, including inadequate safety reviews and poor operating quality assurance, that allowed the use of open flames in the electrical control system of the operating

Browns Ferry Unit; (c) defective installation of electrical control cables that allowed cables controlling redundant equipment to be crowded too close together so that they could be destroyed simultaneously by a small fire; (d) lack of effective emergency procedures for detecting and promptly extinguishing fires; (e) construction and operation of the plant in flagrant violation of important safety requirements.

The report concluded that "the AEC, whose interests over the years had merged with those of the industry, . . . shared TVA's enthusiasm for the project and gave less than adequate attention to . . . safety problems. . . . In particular, the AEC ignored years of warnings from its safety reviewers and inspectors about dangers of electrical cable fires arising from poor control of combustible materials, inadequate fire prevention programs and poor physical separation of redundant circuitry. According to its own records, the AEC was repeatedly warned during 1969–1974 about the problem of 'catastrophic' fires. . . . Yet the AEC, aware of the potentially unsafe conditions . . . allowed the plant to go into full power commercial operation. . . ."

Atom-waste mixture explodes. A chemical explosion in a 13-liter container of radioactive wastes at the Hanford Nuclear Reservation in Washington State Aug. 30, 1976 injured a workman and contaminated him and nine others with radioactivity.

Eight of them, including two nurses, were decontaminated shortly after the explosion. Two were hospitalized for radiation observation. One of them, Harold McCluskey, 64, was working with a "glove-box chamber," manipulating the material inside it, when the explosion occurred. The blast shattered the box's plexiglas windows, showering McCluskey with shards. Also held for observation was Marvin E. Klundt, 43, who was exposed to radiation when he went to McCluskey's aid.

(A "glove-box chamber" was a device with holes cut into the sides and air-tight gloves fitted into the holes. A workman used the gloves to reach inside the chamber and manipulate the radioactive material inside.)

A medical report Sept. 1 said that Mc-Cluskey had received an excessive radiation dose and had problems with his vision, nitric-acid burns on his skin and possible internal radiation contamination.

The explosion occurred in a small building used by the Atlantic Richfield Hanford Co. for radioactive-waste recovery. The waste was produced by nuclear reactors. The material being recovered was americium, a radioactive element used in petroleum exploration to measure ground heat and in medical processes as a source of radiation.

The mixture that blew up was said to be at least 10 years old. George Stocking, president of Atlantic Richfield Hanford Co., reported Aug. 30 that the explosion was caused by a chemical reaction and that, "as far as we know, no radioactivity was released into the atmosphere, but some material was tracked out of the room but remained confined to the building."

Orders to deactivate one of the two nuclear reactors at the Hanford Nuclear Reservation had been suspended Feb. 4, 1971 by President Richard M. Nixon.

It had been reported at the time that the reactor produced plutonium for nuclear weapons and electric power for the Washington Public Power Supply System. The suspension order came after hearings by the Joint Congressional Committee on Atomic Energy Feb. 4 at which Sen. Henry M. Jackson (D, Wash.) testified that the reactor's closure would severely affect power supplies and employment in the Hanford area.

An unidentified Nixon Administration official, quoted by the New York Times Feb. 6, called the reactor "unreliable and a possible safety hazard."

Plant & Equipment Problems

AEC curbs NY A-plant. The AEC reported July 27, 1972 that it had ordered a power plant in Rochester, N.Y. to reduce power to 83% of capacity in June, after damage to some fuel rods was found.

The agency reported that about 4% of the rods had been flattened in sections, raising the danger of cracks and "higher levels of gaseous radioactivity in the reactor coolant system." Although three other operating nuclear power plants in the U.S. used the same type of fuel rod, the AEC said none had developed problems.

Michigan breeder reactor closed. The closing of the first reported U.S. operating nuclear breeder reactor power plant at Lagoona Beach, Mich. was announced Nov. 29, 1972. The Enrico Fermi plant was effectively decommissioned in 1974.

The Fermi nuclear reactor had run into continuous operating difficulties in its nine-year history, and was considered obsolete.

Plutonium leak. The AEC said May 14, 1974 that it had detected leakage of a highly-toxic plutonium isotope from its plant in Miamisburg, Ohio, a suburb of Dayton. The plant produced nuclear power supplies for satellites and other spacecraft.

Based on "preliminary samples," the AEC said, the leakage presented no health problems because it had been found deep in mud in the Erie Canal near the plant, not in air or vegetation.

Power plant cutbacks ordered. The AEC Aug. 24, 1973 ordered 10 nuclear power plants in seven states to cut back power levels 5%–25% pending studies of possible safety hazards.

The AEC said the precaution was taken because of the discovery of shrinkage in uranium oxide pellets in reactor fuel rods. With the shrinkage, the heat of the atomic process was not efficiently transferred to cooling water, narrowing the safety margin in case of cooling system failure. The shrinkage could also cause collapse of fuel rods.

The cutbacks would be in effect until the commission could evaluate new data from General Electric Co., manufacturer of the reactors.

Power plant 'events' reported. The AEC said May 28, 1974 that 861 "abnormal events" had occurred at the U.S.' 42 nuclear power plants during 1973. None of the incidents resulted directly in health hazards, the agency said, but 371 were potentially hazardous. Twelve of the incidents involved release of radioactivity at rates above permissible limits beyond plant site boundaries, although the total amount released was said to be within safety limits.

The report said the incidents included loss of power supply, failures of electronic monitoring equipment and cooling system leaks. Each of the 42 plants recorded at least one "event."

Cracks bring BW reactors inspections. The NRC, a successor to the AEC, said March 7, 1975 that 21 of the nation's 23 boiling water reactors had passed a safety inspection ordered after cracks were discovered in the emergency cooling system of an atomic reactor operated by the Commonwealth Edison Co. near Morris, Ill.

An additional crack had been found at the Morris plant and one other reactor inspection, also at Morris, was deferred.

At a Congressional hearing Feb. 5, agency officials admitted the cause of the cracks was unknown but considered the shutdown of the plants for the inspection "appropriately prudent." Several senators at the hearing questioned the reliability of safety inspections conducted by public utilities operating the plants.

Another witness, Daniel Ford, executive director of the Union of Concerned Scientists, a public interest environmental group, disagreed. He questioned the fitness of the regulatory licensing procedures, in view of the emergency shutdowns, as well as the safety and reliability of reactors themselves.

Documents made public by the NRC at Ford's request disclosed concern within the agency about the nuclear reactor safety issue. The New York Times published March 9 a report on a policy study by the NRC's Edwin G. Triner concluding that utilities owning most of the nuclear reactors were not sufficiently concerned about safety and performance. A N.Y. Times report published April 6, also based on the newly released material, revealed that several NRC scientists and technicians viewed the test for the cracks in the reactors as unreliable.

Radiation leaks shut Md. reactor. The Baltimore Gas & Electric Co.'s nuclear power plant in Calvert Cliffs, Md. said Jan. 25, 1975 that it would close Feb. 1 because of unacceptable radiation leaks from the door to the reactor container. The NRC had reported Jan. 24 that the emissions were from three to 10 times above federal standards.

Geologic fault at reactor site. The Virginia Electric & Power Co. (VEPCO) informed the AEC May 17, 1973 that a geologic fault had been discovered during site preparation for four nuclear power reactors planned at the North Anna Power Station in Louisa County, Va. After discovery of the fault, extensive hearings were held on the question of whether the facilities could be built and operated safely. Experts from AEC and the U.S. Geological Survey testified that the fault was not "capable"—inactive and not of safety significance.

The AEC also ordered, in May 1974, a hearing on allegations that VEPCO had made false statements on seismic conditions at the site. The NRC's licensing board Sept. 11, 1975 imposed a $60,000 fine on the utility. The NRC said that it found VEPCO's "delay in informing the board and the explanations given for that delay unacceptable."

VEPCO paid the fine imposed by the commission Feb. 23, 1977. The fine was the third largest ever levied in civil actions by the NRC or its predecessor, the AEC. VEPCO had received the two larger ones also.

Vt. power plant shut. The Vermont Yankee Nuclear Power Corp. closed its 540,000-kilowatt nuclear power plant in Vernon, Vt. Jan. 26, 1976 "as a safety precaution," the company announced Feb.

2. The problem reportedly involved the safety container around the reactor and a possible design flaw. A breakdown of the safety barrier and escape of the radioactive core materials was considered the worst possible nuclear accident.

NRC reports "abnormal occurrences" for fiscal 1977. During fiscal year 1977, the NRC reported a number of events that were described as abnormal occurrences. By law—the Energy Reorganization Act of 1974—the NRC is required to submit to Congress a report listing any "abnormal occurrence" at or associated with any nuclear facility in the U.S. An abnormal occurrence is an "unscheduled incident or event which the commission determines is significant . . . to the public health or safety."

Among the events reported by the NRC:

As the Millstone Power Station was being shut down for maintenance July 5, 1976, the reactor inadvertently "tripped"—automatically shut down. Several motors, not safety-related, failed to start up as they should have under such circumstances. A July 21 incident caused the loss of emergency and auxiliary power for about five minutes. The cause of these incidents were said to have been design defects for which corrective action was later taken.

The Florida Power & Light Co. of Miami, Fla. in 1976 reported unpredicted behavior affecting the nuclear core distribution in its St. Lucie Plant. The NRC confirmed that the incident had been the result of a manufacturer's defect in the reactor rods that are used to help distribute power in the core.

The Millstone Nuclear Power Station, Unit 1 in Connecticut was shut Nov. 12, 1976 for refueling at the time an unplanned reactor criticality—the state of a reactor when it is just sustaining a chain reaction—took place in the reactor core. There was no risk to the general public from this episode. There were however, plant personnel on the refueling floor whose safety was jeopardized. Following the mishap, the NRC halted fueling operations until corrective measures had been taken. The NRC said that "the event was caused by a combination of personnel error, procedural inadequacies and a failure in administrative control." Subsequent enforcement action by the NRC included the imposition of a $15,000 fine.

A breach in the security system was made April 19, 1977 at the Fort St. Vrain Nuclear Generating Station in Colorado when an NRC inspector passed through several guarded checkpoints and gained access to the reactor's control before he was stopped. The NRC said, had this penetration been achieved by a saboteur, he may have been able to carry out his intentions. Following this incident, the NRC issued more stringent security regulations in February 1977.

After an inspection of operations at the Ohmart Corp. of Cincinnati, Ohio March 30-31, 1977 the NRC had concluded that the company's practices threatened the health and safety of its employees. The firm manufactures and distributes gauging devices containing radioactive sources. The inspection had revealed that employees handling radioactive sources had suffered exposures in excess of the regulatory limits. The over-exposures was judged to have been caused by the company's failure to exercise sufficient management control. The NRC ordered the firm to train its employees in the safe handling of radioactive sources and improve handling techniques.

International Developments

West Germany: *2 die in plant accident.*
Two workmen were killed Nov. 19, 1975 at a nuclear plant at Gundremmingen, West Germany after suffering burns caused by escaping steam. The deaths, reportedly had nothing to do with radioactivity, were the first deaths at a nuclear plant in the country. They occurred as the men were tightening joints of a pipe that took steam from the reactor core to the plant's turbines.

India: *radioactive danger denied.*
Atomic Power Authority chairman J. C. Shah denied U.S. reports that leakage

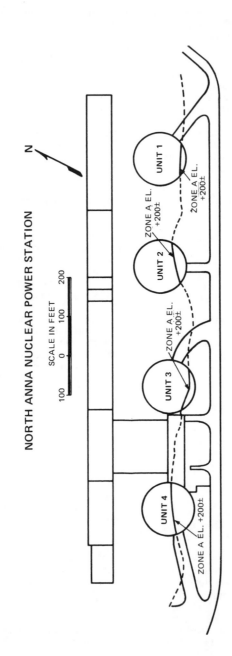

FAULT AT A-PLANT FOUND NOT DANGEROUS

NORTH ANNA NUCLEAR POWER STATION

Chlorite seams observed in 1970–71 in connection with construction of Units 1 and 2 of the North Anna Nuclear Power Station were found in 1973 to continue into the site of Units 3 and 4. Some of the seams were later determined to be faults in the excavations for Units 3 and 4. The drawing traces the path of the fault of primary concern—Zone A—through the site of the four containment buildings at elevation +200±. The fault was determined to be "not capable," that is, inactive and not of safety significance. Source: NRC Annual Report (1977) from the Tarapur atomic power plant near Bombay, India had spread radioactivity along the seashore and affected fish-eating residents of the area, it was reported Feb. 7, 1976. Shah said "there is bound to be some radioactive fallout" at an atomic plant, but insisted that "the authorities have kept it at a level lower than tolerable and there is no concern at all."

Shah's statement was made in response to testimony given Jan. 30 by U.S. Sen. Alan Cranston (D, Calif.) at a hearing of the Senate Government's Operations Committee. Cranston cited an official of the U.S. AEC who had visited the Tarapur plant in 1972.

USSR: *A-disaster reported.* Zhores Medvedev, an exiled Soviet scientist who was living in Britain, Nov. 6, 1976 described an explosion of nuclear waste material in the Soviet Union's Ural Mountains in 1958 that he said left hundreds of acres contaminated and afflicted thousands of persons with radiation sickness. In an article published in the British scientific weekly New Scientist, Medvedev said that reactor wastes stored underground near the city of Blagoveshensk on the Asian side of the mountain chain exploded "like a volcano." He said that strong winds carried radioactivity hundreds of miles away.

Sir John Hill, chairman of the United Kingdom Atomic Energy Authority, Nov. 7 called Medvedev's account "science fiction" and "a piece of nonsense." Medvedev Nov. 8 retorted that he had learned of the incident from a scientist for whom he had been working at the time. The scientist, Medvedev said, had set up stations to study the radiation effects on plant and animal life in the area.

U.S. experts cited in the Nov. 17 Washington Post did not discount Medvedev's story, but suggested that the blast might have happened in 1957 and might have been triggered by severe earthquakes which occurred that year. They said that nuclear wastes were stored in Troitsk, on the European side of the Urals and not in Blagoveshensk. They said that Medvedev might have been mistaken about the loca-

tion because as a biochemist he would not have had detailed information about the Soviet nuclear program. The U.S. experts said that the storage tanks for the reactor wastes had probably been buried no more than 20 feet apart, so that if one exploded, pieces of the steel hull could easily pierce neighboring tanks. (By comparison, they said, U.S. nuclear-waste storage tanks were built of double-walled stainless steel and were buried at least 500 feet apart.)

Leo Tumerman, an exiled scientist living in Israel, said Dec. 8 that he had seen the results of the disaster during a 1961 trip from Sverdlovsk to Chelyabinsk. About 65 miles from Sverdlovsk, he said, he drove through an area where signs warned motorists to close their windows and drive as fast as possible because the area was contaminated. He described "empty land . . . Only chimneys . . . for many hundreds of square kilometers, useless and unproductive for . . . maybe hundreds of years." He and others blamed "Soviet officials who were negligent and careless in storing nuclear wastes."

France: *A-plant leaks radioactive gas.* Radioactive gas leaked July 1, 1977 from an atomic processing plant near Pierrelatte in France. The accident occurred when a valve on a uranium hexafluoride container was broken by a worker. Nine employes who inhaled the gas were given medical attention. A small cloud contaminated part of the factory and the surrounding area, but, according to officials, high winds scattered the cloud before serious contamination took place.

Another accident at the same plant two months earlier had contaminated the staff's drinking water.

Uranium hexafluoride, a highly toxic gas, was used to produce enriched uranium. The Pierrelatte plant produced about one quarter of the world's supply of uranium hexafluoride.

Belgium: *nuclear plant leak disputed.* Environmentalists clashed with Belgian nuclear authorities Jan. 25, 1978 at public hearings on nuclear power that were being held by the European Community Commission in Brussels. Environmentalists from the Belgian branch of the Friends of the Earth claimed that 80 workers had been contaminated Jan. 13 at the Tihange pressurized-water reactor when radioactive iodine gas 131 escaped.

Robert van Damme, a director of Electrobel of France and Intercom of Belgium, appeared at the hearings to refute the environmentalists' claim. He admitted that a gas leak had occurred at the facility, but said that none of the workers affected had required hospitalization. He was supported in his statement by Dr. Edward Teller, the U.S. scientist known as the "father of the hydrogen bomb." Teller testified at the hearings to encourage the development of nuclear energy sources.

Britain: *plutonium facility shut.* Great Britain reported that it had closed its Atomic Weapons Research Establishment at Aldermaston Aug. 24, 1978 after members of the staff were discovered to have excessive amounts of plutonium in their lungs.

The Defense Ministry had become aware of the problem earlier in the month when three women who worked in the Aldermaston laundry, where contaminated clothing was washed, were found to have high plutonium levels in their lungs. The ministry then announced Aug. 16 that nine men who had been preparing contaminated waste for disposal and working in other areas involving plutonium handling were also contaminated.

The high plutonium levels were discovered through a new radiation monitoring technique called "whole body" monitoring. Officials were not sure whether the high levels discovered in the 12 workers were the result of increased exposure or the improved detection technique.

According to press reports Aug. 18, the affected workers were not expected to become ill from their exposure to plutonium since the amounts involved were small.

John Pardoe, the Liberal Party's spokesman on finance, said Aug. 17 that the contamination of workers reported at Aldermaston proved that exponents of Britain's nuclear energy program were wrong when they dismissed the possibility of accidents occurring at nuclear facilities.

Water Protection

Environmental Policy Act

The passage of the Environmental Policy Act of 1969 (NEPA) gave the Nuclear Regulatory Commission a broader environmental responsibility in monitoring the development of atomic energy.

AEC procedures violate NEPA, new rules issued. The U.S. Court of Appeals for the District of Columbia ruled July 23, 1971 that the AEC, by considering only radiation hazards in the licensing of nuclear power plants and not other environmental effects such as thermal pollution, violated the 1969 National Environmental Policy Act (NEPA).

The court issued its ruling in an environmental suit seeking to halt construction of the Calvert Cliffs plant on Chesapeak Bay in Maryland.

To conform with the decision, the AEC Sept. 3 announced new rules under which the operating licenses or construction permits of more than 100 nuclear reactor power plants in 21 states would come under review.

The AEC speculated that the new regulations would cause a construction delay for 81 atomic power plants and could result in the closing of five plants already in operation.

Ten plants that were granted provisional operating licenses before the 1970 act also would be reviewed for environmental effects, but the licenses would not be suspended. Seven other facilities, the oldest nuclear power plants in the country, faced no action because their full operating licenses were granted before 1970.

The AEC shifted the burden of proof to the utilities by ordering those affected to submit information within 40 days "showing why his permit or license should not be suspended in whole or in part" pending completion of a review to produce a complete environmental impact statement. It was estimated that adequate environmental safeguards could cost the nation's utilities as much as $25 million.

The AEC announced Nov. 24, 1971 that it would not halt construction on the Calvert Cliffs plant and three others while it conducted reviews of the plant's environmental impact.

An AEC aide said Nov. 24 that each decision would be based on whether further construction would significantly harm the environment or foreclose any design changes the review might find necessary, and on the adverse effect of a delay on costs, power needs and the local economy.

Citing data supplied by the Baltimore Gas & Electric Co., builder of the largely completed Calvert Cliffs plant, the AEC claimed that a delay during the six-month review would add $162 million in construction, finance and other costs, would idle 2,500 workers in a county with 1970 unemployment of 7.4%, endanger future tax revenues for the county, which had already floated $6.5 million in bonds in anticipation, and lower power reserves in the Pennsylvania-New Jersey-Maryland power pool. The AEC also said that further work would not add to the esthetic damage already done to the site.

The AEC stressed that final approval of the plant would only come after the review, which would be completed before reactors were installed. Opponents had warned of ecologically disruptive temperature increases in Chesapeake Bay, since the plant would use over a billion gallons of water daily as a coolant.

Other plants receiving tentative go-aheads were the Yankee station in Lincoln County, Me., the Fitzpatrick plant near Oswego, N.Y. and the Oconee, S.C. power station. Two other plants had previously been approved, while a partial halt had been ordered in construction of a plant near Wilmington, N.C.

Thermal pollution suit settled. The Justice Department announced settlement Sept. 1, 1971 of the first federal suit over thermal pollution. The department had filed suit in 1970 charging the Florida Power and Light Co. with disrupting the ecology of Biscayne Bay by discharge of heated water.

The department suit which had been filed on the recommendation of Interior Secretary Walter J. Hickel alleged that heated water being discharged by two power plants was ruining marine life in the bay, including an area designated by Congress as the Biscayne National Monument. The suit claimed that damage would be even greater when two planned nuclear power plants were installed at the existing sites.

The government had said that the plants drew water from the bay at a rate of almost 550,000 gallons a minute for cooling and condensing purposes; that the water, which left the plant at about 10 to 20 degrees above the normal temperature of the bay, raised the bay water to "temperatures substantially higher than their natural condition"; that 300 acres were damaged in 1968 and 300 more acres in 1969; that when the nuclear reactors were completed, heated water equal in volume to all of the bay would pass through the plant in less than month; and that the nuclear plant intake would disturb the bay bottom and disrupt marine life.

The government had also asked that the company be permanently enjoined against operating "its presently existing fossil-fueled power plants and its nuclear reactors now under construction as to increase the temperature or otherwise adversely affect the quality of the waters of Biscayne Bay or to adversely affect marine life of the Biscayne Bay National Monument and the lands therein."

After the Justice Department agreed to a settlement that would halt thermal pollution by the utility, Attorney General John N. Mitchell announced a consent decree filed by the government under which the utility agreed to build a $30 million system to insure that discharged water would be compatible with the bay water in salinity and temperature. The company used the bay water to cool its generators.

After a hearing Sept. 9, 1971, District Court Judge C. Clyde Atkins told both sides he agreed with the major points in the settlement and would grant approval with adjustments in the wording of the decree.

Florida to curb thermal pollution—The Florida pollution control board Nov. 22 proposed stiff curbs on hot water discharges from electricity generating plants.

The rules would bar discharges that raise temperatures in surrounding waters more than four degrees in winter and 1.5 degrees in summer.

AEC orders N.Y. plant cooling change. In its first regulatory act concerning thermal pollution at any atomic plant, the AEC ordered the Consolidated Edison Co. (Con Ed) Oct. 2, 1972 to install a "closed cycle" cooling system at its Indian Point, N.Y. complex by 1978, in order to end thermal pollution and massive fish kills in the Hudson River.

The commission ordered Consolidated Edison to replace its current system, which drew over a million gallons of water per minute from the Hudson and returned it 20 degrees warmer, with two 400-foot-high cooling chimneys, which would recycle the water. The chimneys were expected to cost as much as $150 million to build and $2.5 million a year to operate.

The utility was granted a five-year delay to alleviate the scarcity of power in the New York area, and to allow the company to retire air polluting plants in New York City.

Con Ed found guilty—Con Ed was ruled guilty in New York Supreme Court Oct. 12, 1972 of violating the state conservation law's prohibition against the taking of fish by drawing off water. The company faced fines of over $1,300,000 for inadvertently killing 130,000 fish between Feb. 22 and Feb. 26 through the Indian Point cooling system.

New York had filed suit in state Supreme Court May 12, 1970. The suit charged Con Ed with "creating serious conditions of thermal and chemical pollution in the Hudson River" and "endangering the ecology." The suit asked that the plant be closed until "suitable methods" to protect the Hudson were found and sought $5 million in damages for fish kills resulting from operation of the plant.

Fish kills from the Indian Point plant were reported as long ago as 1966, and two separate fish kills were reported in 1970. In January, an estimated 150,000 fish were killed, requiring a change in a filtering device designed to screen fish

out of the plant's water intake system. In March, 120,000 fish were killed by what officials said were "unknown reasons."

Consolidated Edison officials expressed concern over the suit because the Indian Point plant was counted on to produce "about 3% of our total power generating capacity."

Quad Cities A-plant testing licenses granted. The AEC issued testing licenses March 31, 1972 for the Quad Cities atomic power plant at Cordova, Ill. after the utilities agreed to construct a four-mile $30 million cooling system to prevent the thermal pollution of the Mississippi River. A district court Dec. 13, 1971 had barred the licenses until the AEC issued a complete environmental impact statement called for by the NEPA.

District Court Judge Barrington D. Parker interpreted the National Environmental Policy Act's provision for impact statements as a "mandatory obligation not within the discretion of the agency," which "must be performed in all circumstances prior to the taking of the agency action."

The injunction had been sought by the Illinois attorney general and the Izaak Walton League of America, a conservation group, amid reports that the AEC had planned to grant a permit to the Commonwealth Edison Co. and the Iowa-Illinois Gas and Electric Co. to run the plant at 50% capacity through March 1972. Theodore Pankowski of the Izaak Walton League said the ruling "takes the Calvert Cliffs decision and pushed it one step forward." He said the court ruling might prevent an alleged trend by federal units to avoid researching their own impact statements, instead accepting those of applicants.

Judge Parker's decision cited expected temperature rises in river water used as a coolant for the plant, which might have adverse effects on fish reproduction.

AEC urges interim licenses. The AEC, backed by top Administration environment officials, asked Congress March 16,

1972 for an amendment to the National Environmental Policy Act to permit the AEC to grant interim emergency operating licenses to new nuclear power plants prior to filing complete environmental impact statements.

AEC Commissioner James R. Schlesinger told the Joint Committee on Atomic Energy that without emergency licenses, five completed nuclear plants would be unable to meet possible 1972 and 1973 power shortages in New York City, Illinois, Iowa, Michigan and Wisconsin.

Under the proposed amendment, the AEC would have the power through June 1973 to issue licenses without public hearings and without submitting statements to the Council on Environmental Quality. After July 1, 1973 no new interim licenses would be issued, but any license could be continued if the AEC declared the emergency still in effect.

Russell E. Train, chairman of the Council on Environmental Quality, and William D. Ruckelshaus, administrator of the Environmental Protection Agency, testified in favor of the amendment, leading some congressmen to charge that the Administration was planning to cripple the environmental act.

The attempt to amend the NEPA was tabled July 19 by the Senate Interior Committee, killing chances for 1972 passage.

Some senators, including Republican-Conservative James L. Buckley (N.Y.) feared that the amendment would set a precedent endangering the entire environment act, while others said the Atomic Energy Commission (AEC) had not specified any region which would suffer a power shortage this summer from environmental delays in licensing completed plants.

Temporary permits OKd—The Senate approved a House-passed measure 80–0 to grant temporary operating permits for nuclear power plants whose permanent licenses were under challenge, it was reported May 19.

The bill, designed to prevent possible power blackouts or brownouts in 1972, had been passed by the House May 3.

The AEC would be empowered to grant the permits if it determined that safe operation were possible, until Oct. 30, 1973. About 12–15 plants nearing completion would be affected, whose permanent licenses were in some cases being challenged on environmental grounds.

The bill had been revised by the Joint Atomic Energy Committee, after criticism by environmentalists, to require a public hearing and a detailed environmental impact statement before even the temporary permit could be granted.

Radioactive wastes dumping banned. Congress passed and sent to the White House Oct. 13, 1972 a bill that would ban the dumping by U.S. citizens of any radiological warfare agents or any highly radioactive debris in oceans beyond the territorial limit.

President Nixon approved the bill Oct. 28.

U.S. A-sub discharge accident. The U.S. Navy said Jan. 5, 1972 that the accidental discharge of more than 500 gallons of reactor coolant water into the Thames River at New London, Conn. had caused "no increase in the radioactivity of the environment." A nuclear-powered submarine had discharged the water Dec. 29, 1971 but the Navy had not at first disclosed the incident.

Connecticut Gov. Thomas J. Meskill asked the Navy and AEC Jan. 6 to set up a liaison system to inform state authorities of future accidents.

N.Y. State A-plant abandoned. The New York State Electric & Gas Corp. said July 14, 1973 that it had abandoned plans to build a nuclear plant on the shore of Cayuga Lake near Ithaca. Environmental groups had charged that heated water discharged from the plant would damage the ecology of the lake. Company President William A. Lyons said the expected legal delaying tactics by opponents of the plant had forced the decision. The company was preparing plans to build a conventional, coal-fired plant on the same site.

Nuclear Wastes

U.S. A-Waste Programs & Problems

Federal nuclear waste policies under attack. In mid-1973 AEC Chairman Dixy Lee Ray ordered a re-evaluation of nuclear waste management programs after a controversy developed over leaks from waste storage tanks at the AEC facility at Hanford, Wash., it was reported Aug. 28. Dr. Ray conceded that radioactive waste management had at times been "sloppy" and "negligent," but she said public health had not been endangered.

The 115,000-gallon Hanford leak had been confirmed June 8. The AEC reported July 31 that an investigation had found questionable waste practices being followed by a private contractor, the Atlantic Richfield Hanford Co. The AEC said the leak had gone undiscovered for six weeks because a company supervisor had not read reports showing waste levels dropping steadily in the holding tank.

The AEC reported, however, that the liquid waste had stabilized at a point underground where there would be no hazard to ground water.

Four environmentalist groups filed suit in federal district court in Spokane Aug. 1 seeking to prevent resumption of the Hanford operation after repair of the leaks.

The suit accused the AEC of failing to file the required environmental impact statement for the Hanford operation and of violating radioactive wastes sections of the Atomic Energy Act of 1954.

The plaintiffs alleged that, in addition to accidental leaks of high-level wastes, low and intermediate-level wastes had been deliberately disposed of in the soil, endangering water supplies. The AEC said the high-level wastes had been contained and the other wastes posed no dangers. The suit was dismissed Aug. 17 after the AEC agreed to file an environmental impact statement.

Nuclear-power critic Ralph Nader released a draft report by a nuclear energy expert Sept. 7, 1976 concluding that "a major radioactive waste problem already exists" in the U.S. The report was prepared by Mason Willrich for the Energy Research and Development Administration (an agency that was one of the AEC's successors).

The report said that the federal government's past handling of radioactive material had been "marred in a sufficient number of instances to be a cause for concern" and that the system of storing wastes soon "will be unworkable." It said further that the escape of radioactive material into the air and water would "constitute a radiological hazard for hundreds of thousands, perhaps millions of years."

It was estimated that 75 million gallons of high-level radioactive waste and 51 million cubic feet of low-level waste were stored at nine different sites in the U.S. The federal military reactors were said to be producing 7.5 million gallons of liquid high-level waste a year and the commercial reactors were expected to produce a total of 60 million gallons of such waste by the year 2,000.

Federal production of low-level radioactive waste was put at 1.3 million cubic feet a year, the commercial production was estimated at a total of 50 million cubic feet by 2,000.

At the Hanford site, the government's major storage area, 18 leaks had resulted in losses of 430,000 gallons of high-level wastes into the surrounding soil, the Willrich report said.

The problems of the removal or stabilization of the Hanford wastes was studied by a National Academy of Sciences committee. Based on a report sponsored by the Ford Foundation in 1977, the committee considered a plan to permanently cover the five-square-mile site with concrete. The report asserted that the area is so "badly contaminated that the land may never be cleaned up." It warned that "whether these wastes really can be sealed off for hundreds of thousands of years and the land removed from man's use for this period is far from clear."

Another waste disposal problem that had not been solved, the Ford study said, was at the Nuclear Fuel Service's commercial reprocessing plant at West Valley, N.Y. The West Valley plant had been closed in 1971 because it could not meet federal radiation standards. Over 600,000 gallons of waste were temporarily stored in a single 750,000-gallon carbon-steel tank and another 12,000 gallons in a 15,000-gallon stainless steel tank; other studies indicated, however, that waste treated in this manner cannot remain harmless permanently. "The past history of management practices and the problems left by them . . . suggest the need for greater care in future decisions . . . ," the Ford study said.

Atom-waste mixture explodes—A chemical explosion in a 13-liter container

of radioactive wastes at the Hanford Nuclear Reservation in Washington State Aug. 30, 1976 injured a workman and contaminated him and nine others with radioactivity.

Eight of them, including two nurses, were decontaminated shortly after the explosion. Two were hospitalized for radiation observation. One of them, Harold McCluskey, 64, was working with a "glove-box chamber," manipulating the material inside it, when the explosion occurred. The blast shattered the box's plexiglas windows, showering McCluskey with shards. Also held for observation was Marvin E. Klundt, 43, who was exposed to radiation when he went to McCluskey's aid.

(A "glove-box chamber" is a device with holes cut into the sides and air-tight gloves fitted into the holes. A workman uses the gloves to reach inside the chamber and manipulate the radioactive material inside.)

A medical report Sept. 1 said that McCluskey had received an excessive radiation dose and had problems with his vision, nitric-acid burns on his skin and possible internal radiation contamination.

The explosion occurred in a small building used by the Atlantic Richfield Hanford Co. for radioactive-waste recovery. The waste was produced by nuclear reactors. The material being recovered was americium, a radioactive element used in petroleum exploration to measure ground heat and in medical processes as a source of radiation.

The mixture that blew up was said to be at least 10 years old. George Stocking, president of Atlantic Richfield Hanford Co., reported Aug. 30 that the explosion was caused by a chemical reaction and that, "as far as we know, no radioactivity was released into the atmosphere, but some material was tracked out of the room but remained confined to the building."

Senate debates waste disposal policy. Sen. Jennings Randolph (D, W.Va.) told the Senate Sept. 18, 1978 that "[a]lthough waste disposal is the last step in the nuclear fuel cycle, it is the primary nuclear issue facing us today." Randolph said that "despite continuous planning efforts and studies by several federal agencies, I have seen no indication of a proven technology or definitive long-term plan for the disposal of radioactive wastes."

Randolph continued:

"The waste byproducts of nuclear plants are radioactive and deadly. Each of the 70 nuclear plants licensed for operation produces an average 30 tons of radioactive waste per year. Additionally 88 plants have received construction permits and are in different stages of completion. This means that soon the country will have to deal with 5,000 tons of nuclear waste per year. Some of it, the fission products, becomes harmless in a hundred years or so, but it also contains about 2½ tons of plutonium and other elements which must be isolated for at least 250,000 years.

"If nuclear power does become essential to our energy independence we must be able to contain the toxic wastes as we contain those from other human activities. But, at present, no permanent disposal facilities for reactor wastes exist.

"The spent fuel is stored under water at the reactor site for a few years. Throughout the United States, the temporary storage pools for radioactive wastes located on the site of nuclear power plants are filled to capacity, or soon will be. Further, there are persistent reports of problems with the tanks which store high level wastes generated by defense-related activities.

" The situation is rapidly becoming intolerable. We cannot expect States and communities to permit the continued construction of nuclear power plants without provision for the safe disposal of the deadly wastes they generate. Gov. Hugh Carey recently announced that he will seek to bar any new nuclear plants in New York State because of the lack of waste disposal facilities. California and Maine have passed laws to that effect, and others are likely to follow suit.

"These and similar problems must be addressed as expeditiously as possible. We are now in the 21st year of commercial nuclear power, and there is no solution to the waste disposal problem in sight. Again, we must not talk about speeding up the licensing of nuclear powerplants until we have proved that the wastes can be safely isolated for the long times necessary.

"Nuclear advocates often blame environmental intervention and administrative regulation for delays and plant cancellations in the industry. There were delays or cancellations of 67 nuclear plants in fiscal year 1978, but only 14 of these were attributed to licensing and litigation. On the average, delays for these reasons accounted for only about 5 months of the 12 years required to complete a nuclear plant.

"A key factor behind the nuclear slump has been rising costs. Construction costs from 1967 to 1974 increased 300 percent. Cost overruns are exorbitant. The cost of uranium has gone up nearly 600 percent since 1970, increasing much faster than long-term coal contracts.

"Another factor has been the reduced load growth experienced by the electric utility industry. In the 1950's and 1960's this growth rate averaged 7 percent per year. But since 1973 electric growth has averaged about 3 percent per year. Consequently, utilities have delayed or canceled many new powerplants, especially the capital-intensive nuclear plants.

" Ultimately, a decision to continue with nuclear power must rest upon a determination that the technology at every stage of the fuel cycle is safe and reliable.

" So far the administration has not even produced a proposal for action, much less built a facility for radioactive waste disposal. In fact, the major recommendation of a February 1978, study of this matter by the Department of Energy (the Deutch report) was for a further study.

"It is essential that the Federal Government take the lead—and do so immediately—in developing a realistic and workable waste disposal program.... · "

Nuclear waste storage procedures criticized. Sen. Frank Church (D, Ida.) had charged March 6, 1970 that the AEC had failed to release a 1966 report by a panel of the National Academy of Sciences (NAS) critical of AEC procedures for storing radioactive waste.

The report, made available to the press March 6, said that the four major AEC plants where waste was stored were located in poor geological areas for such uses, that the practice of storing waste in the ground could pose a hazard through build-up, and that there was no uniform standard among the plants for determining the degree of radioactivity of waste materials. The panel had said that it saw no immediate danger in the way the AEC was storing or disposing of nuclear wastes.

John Erlewine, AEC assistant general manager for operations, was reported March 6 to have said that the report was only one of many not published by the AEC. He also said that while the AEC was putting waste into the ground it did not necessarily intend to leave it there and its operations were within the standards laid down by the Federal Radiation Council.

Four government agencies later told the AEC it was careless in the way it disposed of radioactive wastes, according to the Washington Post March 13.

The Post cited a report made by the Bureau of Radiological Health, the Bureau of Sport Fisheries and Wildlife, the U.S. Geological Survey and the Federal Water Pollution Control Administration. The agencies recommended that the AEC keep at least two feet of clay or gravel between atomic burial pits and the exposed basalt formations below (the AEC then required no minimum protective barrier) and that the commission take steps to keep atomic burial sites free of flood waters.

H. Peter Metzger, in his book, *The Atomic Establishment* (1972), said that the NAS report called "attention to the fact that the AEC's waste-disposal operations in Idaho 'are conducted over one of the largest of the country's remaining reserves of pure fresh water.' " It added "that with water movement in [the] basalt 'as high as 15 million gallons per day per foot,' the radioactive waste would take only fifty to sixty years to reach the springs in the Snake River Canyon" where the world's largest trout farm is located. A year after the NAS report was made public, Metzger said, the AEC announced that the "buried waste in Idaho would be removed."

AEC's Kansas A-waste plan scored—A 1970 study by the Kansas Geological Society, published Feb. 16, 1971, criticized the AEC's plans for burying radioactive wastes in salt mines at Lyons, Kansas. The report said the AEC had "exhibited remarkably little interest" in studying the effects of radiation and heat on the salt, that plans for transporting the radioactive waste across the state were "completely inadequate" and that emergency plans "do not exist at all." The study said that the salt mine selected by the AEC was the most dangerous of several possible sites in the state. The New York Times reported March 17 that the AEC had assured opponents of the project in Kansas that the dumping scheme would be abandoned if safety problems arose.

Metzger, in *The Atomic Establishment,* said that the AEC, after spending more than $100 million in a 15-year study of salt-bed disposal, was "convinced that the Lyons site [was] equal or superior to . . . others." Salt beds are primarily chosen for waste disposal because they are generally found to have been geologically stable for millions of years. And, as Metzger said, "are among the least likely place for water to exist underground." If water entered the mines and came into contact with the wastes, the wastes—"red-hot and well over one thousand degrees Fahrenheit—would vaporize the water instantly and violently expel steam, water salt and waste particles throughout the mine" and into the atmosphere.

A state geological study said, however, that "not enough is known about the underground water there. A letter sent to the AEC by a spokesman for the American Salt Co. "expressed concern about the presence of water and [warned] that his company had been injecting water into the formation for fifty years; as we remove salt, the water replaces the salt." A later AEC report "revealed that tunnels of the American Salt Company's mine . . . [came] as close as five hundred yards to the . . . proposed atom dump." Metzger said that by September 1971, because of "certain additional information, . . . [the AEC] would have to consider possible alternative sites to Lyons, Kansas."

Alternative sites considered. In the face of opposition to its plans to bury radioactive wastes underground, the AEC announced May 19, 1972 that it would build a huge steel and concrete waste storage facility above ground or just below the surface.

The facility was to be ready by 1979 or 1980, at an estimated cost of $100 million. In the meantime, the AEC said it would continue research on burying the waste products of power plant fission reactions in abandoned salt mines or other underground formations.

The 1966 report by the National Academy of Sciences had advised the underground alternative.

In its continued search for an underground waste disposal site, the AEC began investigating the salt beds of southeastern New Mexico. The site chosen, the Waste Isolation Pilot Plant (WIPP) near Carlsbad was, however, not without its problems. Philip M. Boffey said in an article published in *Science* (Oct. 24, 1975) that the Sandia Laboratories, of Albuquerque, New Mexico, which was conducting tests at the site, reported that one of the drilled test holes "unexpectedly hit a big pocket of brine . . . about 200 feet below the . . . proposed waste disposal facility. . . . Dissolved in the brine were such gases as hydrogen sulfide, which is toxic, and methane, which is explosive." Boffey said that the gases "could pose a safety hazard for workers building the facility or operating it" and that "the presence of the brine solution may indicate that fluid have been migrating underground, thereby threatening the integrity of the site." However, William Armstrong, a nuclear waste engineer, said that "the problems [are] . . . much less severe than at Lyons."

GAO report: *A-waste disposal problems*—Sen. Charles M. Mathias Jr. (D, Md.) told the Senate Oct. 10, 1977 that for over 30 years the U.S. had "limped along without a rational," safe waste management program and that it was "time to remedy this deficiency now."

Mathias inserted in the Congressional Record a 17-page digest of a report to Congress by the controller general on nuclear waste disposal problems.

According to Mathias' summary:

Growth of nuclear power in the United States is threatened by the problem of how to safely dispose of radioactive waste potentially dangerous to human life. Nuclear power critics, the public, business leaders, and Government officials concur that a solution to the disposal problem is critical to the continued growth of nuclear energy.

Radioactive wastes being highly toxic can damage or destroy living cells, causing cancer and possibly death depending on the quantity and length of time individuals are exposed to them.

GAO found:

Public and political opposition to nuclear waste disposal locations.

Gaps in Federal laws and regulations governing the storage and disposal of nuclear waste.

Geological uncertainties and natural resources tradeoffs encountered when selecting "permanent" disposal locations.

Lack of the Nuclear Regulatory Commission regulatory criteria for orderly waste management operations, such as solidification of waste, designing proper waste containers, and transporting nuclear waste.

Overly optimistic schedules for demonstrating the safety of the Energy Research and Development Administration's proposed waste disposal locations and waste management practices.

Lack of demonstrated technologies for the safe disposal of existing commercial and defense high level waste.

Now that commercial reprocessing of spent fuel has been indefinitely deferred, finding solutions to problems in storing and/or disposing of nuclear spent fuel will become a top priority matter.

Nearly all operations that produce or use nuclear materials generate radioactive waste. Most waste comes from the energy Research and Development Administration's military reactors, commercial nuclear powerplants (spent fuel elements)[1] and Federal and commercial fuel cycle activities—mainly fuel fabrication and reprocessing facilities.

This report discusses (1) high level waste, (2) transuranic contaminated waste, and (3) reactor spent fuel. All these materials

---

[1] Spent fuel has not yet been defined by the Nuclear Regulatory Commission as high level waste and may not be, because of its potential value as a source of fuel if reprocessed. This report will consider spent fuel as a "potential" high level waste.

contain transuranic [3] elements which determine to a large extent the degree of long term hazard associated with them because some of these isotopes remain hazardous for hundreds of thousands·of years.

High level waste—has extremely high radioactivity—as much as 10,000 curies [3] per gallon. This waste is characterized by high levels of penetrating radiation, high heat generation rates, and a long toxic life. High level waste is created when reactor spent fuel elements are dissolved in acid to recover unused uranium and plutonium for reuse as nuclear fuel. It is the acid solution remaining that is referred to as high level waste. It contains virtually all the fission products [4] and small amounts of transuranics—such as plutonium—which are not recovered during the reprocessing operations. It is one of the most hazardous and complex of all radioactive wastes to manage.

Transuranic contaminated waste—contains much lower concentrations of radioactivity than high level waste. It is generated by plutonium fuel fabrication and fuel reprocessing facilities and laboratories using transuranic elements. This waste generally consists of absorbent tissues, clothing, gloves, plastic bags, equipment, filters from effluent treatment systems, and fuel hulls which remain after fuel reprocessing.

Spent fuel—contains all the fission and transuranic elements that are found in high level waste and all the uranium and plutonium not used during the nuclear reaction. Spent fuel is characterized by high levels of penetrating radiation, high heat generation rates, and a long toxic life.

Until the 1960s, little effort was made to develop technologies for the long term storage or "permanent" disposal of hazardous radioactive waste. The production of atomic weapons materials and development of commercial nuclear powerplants received the highest priority. Decisions as to the management of military waste were based on short term expediency rather than long term considerations.

Even if all activities which generate radioactive waste were stopped today, the United States still would be faced with a major radioactive waste disposal problem. Radioactive waste has been accumulating for dec-

ades from the Energy Research and Development Administration's military and research development efforts, fuel reprocessing activities, and commercial nuclear powerplant operations.

Today about 71 million gallons of high level waste produced by the Energy Research and Development Administration's plants, which reprocess spent .fuel from production reactors used for the .weapons program, is "temporarily" stored in steel tanks at the Hanford facility in Richland, Washington (50 million gallons) and at the Savannah River facility in Aiken, South Carolina (21 million gallons).

In addition, approximately 3 million gallons are stored in underground tanks and bins at the Idaho National Engineering Laboratory at Idaho Falls, Idaho.

About 13 million cubic feet of transuranic contaminated waste from military and research activities either has been buried or stored retrievably at five principal shallow-land burial sites of the Energy Research and Development Administration. This waste is contaminated with about 1,000 kilograms of plutonium. Some of the 1.3 million cubic feet of radioactive waste generated by the Energy Research and Development Administration each year is contaminated with transuranic elements. About 7 million cubic feet of commercial transuranic contaminated waste is expected to accumulate by the year 2000.

About 600,000 gallons of high level waste has been generated from commercial reprocessing activities and is currently stored at West Valley, New York. Should commercial reprocessing operations resume, estimates are that through the year 2000, an ad-definitely deferred commercial reprocessing of spent fuel.

Commercial reactor spent fuel is accumulating at nuclear powerplants because there are no commercial reprocessors now operating in the United States. Resumption of reprocessing does not seem probable in the near future since President Carter has indefinitelydeferred commercial reprocessingof spent fuel.

If it is finally decided that there will be no further commercial reprocessing, spent fuel elements from existing and future civilian power reactors probably will have to be managed as high level radioactive waste. Meanwhile, nuclear powerplants have had to store their spent fuel in storage pools at the reactor sites. As a result, a backlog of spent fuel is accumulating at the powerplants.

The Energy Research and Development Administration estimates that 1985 is the earliest possible date a geological waste disposal facility or other storage facility to receive spent fuel could be ready. By this time the nuclear industry could be faced with a severe shortage of storage capacity.

Spent fuel and transuranic contaminated waste could be as hazardous to the public

---

[3] Transuranic elements are man-made, long-lived, and extremely toxic. These elements—such as plutonium—are created during the normal nuclear reaction process. They are found in several nuclear fuel cycle operations and are contained in nuclear waste in varying concentrations.

[3] Curie—a measure of the quantity of radioactive material.

[4] Fission products—those isotopes formed during the nuclear reaction process that are not part of the transuranic elements. Some of these isotopes are hazardous for hundreds of years.

health and safety as high level waste. While the Federal Government requires extensive regulatory and public oversight over most nuclear plant operations because they use nuclear materials, the same degree of public protection and independent oversight is not currently required for the storage and/or disposal of these hazardous nuclear materials. This situation needs to be changed.

The Congress provides for an independent review of nuclear activities, including waste disposal by the Nuclear Regulatory Commission. Under the Energy Reorganization Act of 1974, the Commission has specific responsibility for licensing and regulating all Energy Research and Development Administration facilities used for storage of commercial high level waste. It has similar authority for retrievable surface storage facilities and other long term storage facilities for the Energy Research and Development Administration's high level waste. This does not include authority over the agency's facilities which are used for or are part of research and development activities.

The act does not specifically give the Commission licensing authority over the Energy Research and Development Administration's—

research and development facilities or full-scale facilities for the temporary storage and/or long term storage or disposal of commercial and its own transuranic contaminated waste,

facilities for the temporary storage of its own high level waste, or

research and development facilities or full-scale facilities for temporary storage and/or long term storage or disposal of commercial spent fuel.

The Congress either should give the Commission authority over those Energy Research and Development Administration facilities—including research and development facilities—intended for the storage and disposal of its own high level wasted, or provided for other independent oversight and assessment of these facilities. The Congress should also either give the Commission authority over the storage and disposal of transuranic contaminated waste and spent fuel, or provide for an alternate means of independent oversight and review.

After several deades of work, the Atomic Energy Commission did not, and its successor—the Energy Research and Development Administration—has not yet: demonstrated acceptable solutions for long term storage and/or disposal of defense- and research-related high level waste, or satisfied the scientific community that present storage sites are suited geologically for long term storage or disposal.

The Energy Research and Development Administration is investigating several alternatives for managing its military and research wastes, including—

immobilizing in place,

solidifying and disposing at Hanford and Savannah River, and

solidifying and shipping to a Federal geological repository.

Before this high level waste can be moved to a repository, however, major questions involving retrievability from its temporary storage tanks at Hanford and Savannah River must be resolved.

The Energy Research and Development Administration does not now have the technological capability to extract all of this waste from the storage tanks. The waste stored at Hanford and Savannah River makes up 94 percent of the total volume of waste. This waste has been converted into a chemical form that may be unsuitable for long term storage.

The Energy Research and Development Administration is testing methods which may make possible the extraction of up to 99 percent of the high level waste from most storage tanks. However, these methods may not work with some older tanks because of their poor condition. The remaining 1 percent of the waste would contain long-lived toxic radionuclides such as plutonium and strontium-90. The costs of extracting and preparing all of the waste for geological disposal are uncertain. Estimates range from $2 billion to $20 billion.

The Energy Research and Development Administration is exploring alternatives for long term storage or disposal of the waste at Hanford and Savannah River. Alternatives include entombment in the existing tanks if the waste cannot be removed, and removal of the waste and burial at the site, either in near-surface facilities or in deep geological formations. These alternatives present still other questions such as the suitability of these sites for geological disposal. Any facility for long term storage or disposal of the waste at these sites will require licensing by the Nuclear Regulatory Commission.

Since a tremendous backlog of spent fuel (potential high level waste) exists at nuclear powerplants because no commercial reprocessors are operating in the United States, utilities are adopting several options to increase their storage capacity. These capacities are being modified at existing reactors and larger storage facilities are being planned for new reactors. Spent fuel shipment to storage pools within a utility's nuclear powerplant system is another plan that utilities are considering.

As of January 1977, utilities operating 36 of the 63 present nuclear reactors haev notified the Commission of their interest to increase storage capacities at their reactor pools by reducing the amount of space between stored fuel elements (compaction).

The safety of such action has been questioned by the Natural Resources Defense Council. In response, the Commission has undertaken a generic environmental impact statement on the storage of fuel elements.

While the statement has not been completed, the Commission has allowed compaction on a case-by-case basis. According to the Commission, before allowing compaction the safety concerns raised by the Natural Resources Defense Council are addressed in each request for increased storage capacity.

According to the Commission staff there are no significant environmental or safety impacts associated with these individual actions. As of January 1977, compaction has been approved for 14 of the 36 reactors.

The Commission has, in fact, justified allowing compaction for utilities which have shown an immediate need for additional storage capacity in order to maintain electrical generating capability. However, some utilities were allowed compaction without demonstrating such an immediate need.

GAO believes that until the Commission completes its generic environmental impact statement, it should limit through license restrictions, the amount of spent fuel that can be put in storage pools to no more than the amounts for which the storage pools were designed and authorized under the initial operating license. Compaction should only be allowed if the utility can prove to the Commission's satisfaction that (1) it would be forced to shut down operations if increased storage at that site was not allowed, and (2) such action would not increase the safety risk to the public or environment. It is of the utmost importance that the Commission complete and issue the generic environmental impact statement as soon as possible so that unanswered questions can be resolved concerning increased fuel storage at reactor pools.

The Energy Research and Development Administration has begun an ambitious program to demonstrate the safety of placing commercial and military wastes in deep geological formations. It is seeking seven sites for facilities in widely separated areas in the country.

The Energy Research and Development Administration has set 1985 as the year for completing two geological disposal facilities for commercial high level and transuranic contaminated wastes and spent fuel (if and when it is defined as a waste). It also plans to complete four more geological disposal facilities for commercial waste between 1987 and 1991.

Furthermore, the Energy Research and Development Administration plans to build a separate disposal facility by 1983 for its own transuranic contaminated waste, generated by military and research activities. At this facility, it intends to have the experimental capability to determine site suitability for high level waste disposal.

One of the potential geological disposal sites which may be used for the 1983 facility is being developed in New Mexico. This facility might eventually be used for routine high level waste storage; however, the

Energy Research and Development Administration has no established date for storing such waste.

The Energy Research and Development Administration's position has been that the New Mexico location is for its transuranic contaminated waste and to provide experimental capability to determine whether or not the site is suitable for high level waste disposal.

Since public and official sentiment in New Mexico appears favorable to a waste disposal facility and the project is further advanced than the commercial waste repository program—which may not have a site ready by 1985—this site may also serve the needs of the commercial nuclear industry by becoming the first commercial waste repository.

Because the President has indefinitely deferred commercial reprocessing of nuclear spent fuel, the Energy Research and Development Administration has decided to initiate a project to store spent fuel in a proposed Surface Unreprocessed Fuel Facility. In the event the President and the Congress ultimately decide against commercial reprocessing, spent fuel—if defined as waste—might have to be disposed of in the geologic repositories. This will affect the six commercial waste repositories currently being planned by the Energy Research and Development Administration.

The six repositories were arrived at mainly to spead nuclear waste regionally throughout the Nation, and minimize any setback to the program should a potential site(s) prove unacceptable. Storage and/or disposal of spent fuel in geological formations requires more acreage than is needed for storage and/or disposal of high level waste.

While the precise number of repositories which will be needed is not known, officials of the Nuclear Regulatory Commission and the Energy Research and Development Administration indicate that three of the size currently being planned may be all that will be needed. In view of the $200 million cost per repository, plus question of excess capacity, public opposition to nuclear waste disposal locations, and security needs, the Energy Research and Development Administration should evaluate the number of repositories currently planned, and justify on a cost-benefit basis, the number they finally believe will be necessary.

The program for commercial radioactive waste repositories which was supported by the Federal Energy Resources Council [5] faces many obstacles. The most serious and critical is public and political opposition to

_____
[5] The Federal Energy Resources Council had the responsibility for coordination of Administration policies and programs relating to energy. The Council participants included: Council on Environmental Quality, Department of Commerce, Department of

waste disposal sites. The success of this program depends to a great extent on whether the Energy Research and Development Administration can demonstrate to the public and elected officials that it has a sound waste management program and that the disposal of radioactive waste in geological risks associated with the storage and/or formations are low.

The Energy Research and Development Administration twice has been unsuccessful in developing potential waste disposal sites because of insufficient attention to the factor of public acceptance—in Kansas and in Michigan.

Other obstacles in the Energy Research and Development Administration's geological waste disposal program include—

geological uncertainties and natural resource tradeoffs,

questionable demonstration time period estimates,

undemonstrated technology for preparing radioactive waste, and

lacking Nuclear Regulatory Commission criteria for orderly waste management operation.

The Energy Research and Development Administration is aware of these obstacles and is addressing them.

Another aspect of the waste repository program which is not, in our opinion, based on realistic appraisals is the goal of building six waste repositories in the stated time period. This goal appears overly optimistic in estimating the time required to identify, study, design, construct and confirm the feasibility of the repositories. Such an unrealistic schedule could further decrease the public's confidence in the Energy Research and Development Administration's waste management program.

RECOMMENDATIONS

To better insure public health and safety the Congress should amend the Energy Reorganization Act of 1974 to provide for independent assessments of the facilities of the Energy Research and Development Administration—including research and development facilities—intended for the temporary storage and/or long term storage or disposal of commercial and its own transuranic contaminated waste; the temporary storage of the Energy Research and Development Administration's high level waste; and the temporary storage and/or long term disposal of commercial spent fuel.

To provide such an independent assessment Congress should adopt one of three alternatives:

Give the Nuclear Regulatory Commission the authority and responsibility for establishing policies, standards, and requirements in cooperation with the Energy Research and Development Administration for carrying out these assessments.

Retain this responsibility and authority within the Energy Research and Development Administration, subject to certain statutory provisions, to insulate the oversight activities.

Authorize the Nuclear Regulatory Commission to assess periodically the Energy Research and Development Administration's facilities and annually report the results to the agency and the Congress.

In testimony before congressional committees, GAO has stated a preference for the first alternative.

GAO also recommends that the Congress closely scrutinize, through the annual authorization and appropriation process, the progress of the Energy Research and Development Administration's program for long term waste management.

The Administrator of the Energy Resarch and Development Administration should:

Proceed to reevaluate the impact that spent fuel storage and/or disposal will have on its commercial repository program.

Reconsider the need for six high level waste repositories in view of disposal requirements through the year 2000 and justify on a cost-benefit basis the number it finally believes will be necessary.

Reevaluate plans for completing the first two repositories by 1985, considering realistically all social, geological, and regulatory obstacles.

Consider the appropriateness of using the New Mexico location also as a commercial waste disposal site, since by 1985 no other facilities may be ready to receive these wastes and public utilities may no longer be able to store them at the reactor sites unless other facilities are constructed. This should be done without sacrificing or impairing the mission of the site to receive Energy Research and Development Administration transuranic contaminated waste.

The Chairman of the Nuclear Regulatory Commission should:

Proceed on a priority basis to complete its waste repository licensing procedures.

Proceed on a priority basis to include in its waste performance criteria, criteria for the storage or disposal of spent fuel.

Complete and issue the generic environmental impact statement on spent fuel as soon as possible, and in the interim, limit through license restrictions the amount of fuel which can be stored in reactor pools to no more than what was originally licensed for, unless the reactor would be forced to shut down operations.

Interior (U.S. Geological Survey), Environmental Protection Agency, Federal Energy Administration, and Energy Research and Development Administration. The Council's function's were recently transferred to the newly created Department of Energy under President Carter's reorganization plan for the Executive Office of the President.

AGENCY COMMENTS AND GAO'S EVALUATION

The Nuclear Regulatory Commission disagreed with GAO's recommendation that pending issuance of the generic environmental impact statement on spent fuel, the amount of spent fuel that can be stored in a reactor pool be restricted to the amount for which it was originally designed and licensed unless the reactor would be forced to shut down. The Nuclear Regulatory Commission cited several operational and procedural reasons for its position.

While GAO does not take exception to the Commission's reasons, it still believes the recommendation should be implemented. The Commission has not fully determined the overall environmental effects from these individual licensing actions nor has it compared these actions to other alternatives for spent fuel storage, such as storage at centralized storage facilities away from nuclear powerplants. Such an assessment is the objective of the generic statement now being prepared by the Nuclear Regulatory Commission. Until this assessment is completed, GAO believes the Commission should restrict the amount of spent fuel to be stored in a reactor pool. To do otherwise may raise public suspicion and concern that the Nuclear Regulatory Commission has made prejudgmental findings on the overall environmental effects of such individual licensing actions, and as such, could possibly cast doubt on the integrity of the generic statement when issued. Furthermore, these individual actions could potentially foreclose the adoption of other storage alternatives that may be as good or better than allowing each utility to increase their storage capacities at the reactor site.

The Nuclear Regulatory Commission did not disagree with the recommendation to the Congress which could broaden its authority over Energy Research and Development Administration waste storage facilities.

The Energy Research and Development Administration generally concurred with the recommendations concerning its activities and stated that work was already underway. Regarding the recommendation to the Congress, the Energy Research and Development Administration agrees that an independent assessment within its organization has merit from the standpoint of assuring the Administrator and the public as to the adequacy of its nuclear operations. However, it does not consider that the first and third alternatives which would place this responsibility within the Nuclear Regulatory Commission are viable since either would, in its view, impose extraordinary burdens on both organizations without commensurate benefit. It believes such a recommendation would be tantamount to requiring its facilities to be licensed by the Nuclear Regulatory Commission. Further, the Energy Research and Development Administration contends that added Nuclear Regulatory Commission ac-

tivities at its facilities could result in the Commission having to acquire expertise they do not now have and which would, to a large extent, be duplicative of the Energy Research and Development Administration. It stated, however, that it has undertaken a comprehensive study to determine how its current assessment activity could be restructured within its organization to provide greater independent assurance to the general public.

GAO doubts that the Energy Research and Development Administration can in fact, structure an organization within itself to independently assess its waste operations without statutory provisions designed to insulate oversight activities from development functions.

Geological Survey on unresolved problems.

U.S. Rep. James Weaver (D, Ore.), discussing the problems of nuclear wastes, inserted in the Congressional Record June 28, 1978 excepts from a statement made by George B. DeBuchananne of the Office of Radio Hydrology of the U.S. Geological Survey.

Weaver said:

" ... In contrast to the confident assertions by Department of Energy officials that the radioactive waste disposal problem is virtually solved, Geological Expert DeBuchananne points to a number of unresolved problems. He notes that there are significant gaps in 'current knowledge concern(ing) the chemical interaction between the host media and the (radioactive) waste.' He notes that effects which could result in the transportation of radioactive waste away from the storage site 'are imperfectly understood (and) are extremely difficult to predict for the intervals of geologic time required for the isolation of the waste.' He further notes that one of the basic problems is that in determining whether the site is a sound site, the site itself is necessarily disturbed by drilling. He goes on to mention that earthquakes are a significant risk in radioactive waste disposal. He concludes that adequate methods for doing (risk evaluations for seismic activity) that are meaningful for long periods of time must be developed to avoid siting problems. In other words, we do not now have the tools to determine whether earthquakes are going to exist. He also notes that we are presently unable to predict future climatic changes which could affect the repositories' acceptability. We do not presently have the technology for understanding the impermeability or permeability of low-permeability sites; that models do not exist which are capable of understanding the ground water formations; and that the technology to seal up the drill holes, mine shafts and other excavations necessary to create a repository do not presently exist. ..."

The following excerpts are from testimony by DeBuchananne April 4 before the Subcommittee on Nuclear Regulation of the Senate Committee on Public Works:

Presently, geologic containment appears to be the most reasonable alternative for radioactive waste isolation if an acceptable geologic repository can be constructed. Because there will always be some residual uncertainty about the basic process of groundwater migration, waste-rock interaction, tectonism, and climatic variability, a conservative approach to the development of these repositories is called for. This approach should involve the development of clear and detailed statements of geologic criteria, the development of adequate techniques for verifying that a particular site meets these criteria, and a plan for long-term monitoring of the repository to ensure that the natural processes that actually occur are those that were anticipated or are at least benign.

One of the greatest gaps in current knowledge concerns the chemical in teraction between the most media and the waste. A significant factor, for example, is related to the generation of heat by the wastes. The highest repository temperatures would occur during the first tens or hundreds of years. The release of thermal energy in the geologic media will result in a variety of mechanical, mineralogical, and hydrologic effects that may strongly influence transport of wastes away from a repository. These effects, some of which are imperfectly understood, are extremely difficult to predict for the intervals of geologic time required for the isolation of the wastes.

The problem of characterizing the environment of a potential repository is amplified by the necessity to minimize disruption of the integrity of the geologic containment media by excessive drilling of exploratory holes in the repository area. Geophysical techniques, particularly downhole techniques, need to be improved to adequately explore and describe the volumes or rock needed for a repository.

Most of the repository sites will be located within the stable part of the North American continent. However, because of the hazard of high-level waste, the risk of seismic activity exposing or releasing the wastes must be evaluated for every area considered for siting. Adequate methods for doing evaluations of this type that are meaningful for long time periods must be developed to avoid siting problems.

One of the major questions that will have to be asked about any proposed repository of radioactive waste will be the effect of future climate on the hydrologic regimen and geomorphic setting. Major climatic oscillations, with periods on the order of tens of thousands of years, have been a feature of global climate for at least the past million years and maybe expected to continue. Therefore, existing paleo-climatological data need to be reviewed to judge the likelihood of the wastes being exposed during a future erosion cycle and/or transported as a result of a change in the hydrologic regimen.

In order for a repository to be acceptable, the host rock must be essentially impermeable. This raises the question of how to measure permeability in a drill hole in rocks of very low permeability. The technology for doing this needs to be developed.

Because circulating ground water is a potential mechanism for transporting wastes to the biosphere, the existing flow pattern must be completely known and models for contaminant transport must be developed. Mathematical models of fluid and solute transport in granular media are available, but flow in fractures, which is the principal avenue in crystalline rocks and shales, is not well understood. In general, further research is needed to develop methods of defining the spatial distribution of hydrologic parameters of rocks. In particular, the occurrence of fractures and the hydraulics of fracture flow should be investigated.

The assessment of containment properties of the rock at any potential high-level waste repository involves considering the potential movement of transuranium elements as well as other radioactive contaminants. There are several aspects of this assessment which need to be fully understood including the following: (1) chemical reactions between the radioactive contaminants, and the natural earth materials through which they are being transported, and (2) the effect of the high-level waste heat on the migration of waste via ground water.

Drill holes, mine shafts, and other rock excavations in the repository are potential pathways that will have to be evaluated in terms of the role they might play in releasing wastes to the biosphere. All repository openings, regardless of depth, will have to be sealed by some, as yet unknown, technology which will in effect return the site, as nearly as possible, to its original condition. In addition, work is needed on the problem of monitoring a repository and the surrounding area for an indefinite time period to detect changes which might signal a loss of containment.

The problem of designing engineered remedial action for the waste migration problem at the existing disposal sites is complicated by a lack of information on the detailed hydrology of the area. The selection of future disposal sites, that will be required as the expansion of nuclear energy continues, will be dependent upon detailed knowledge of the hydrology of the proposed area plus a better understanding of several hydrologic factors that are directly related to the migration of waste.

In the solute transport of waste through a porous media, no model has been developed that will predict the concentration of a solute prior to actual introduction of waste into the aquifer. The theory is well described but actual field tests are lacking. Several studies are presently underway which will help resolve the obstacle.

As noted earlier, the flow of solutes through fractured rocks is not yet adequately understood. All verified transport models assume intergranular flow in the aquifer. However, where flow occurs in a fractured media these models are not applicable. Sound theory relating flow in fractured media to that intergranular flow is presently lacking and must be developed before meaningful field tests can be made.

Theory for chemical reactions among water, waste, and earth materials in the unsaturated zone must be developed. Research on this subject is underway at the present time. After development of adequate theoretical concepts, the next step would be field testing and verification of the theory.

For simple chemical reactions of short duration under conditions of rapid flow, hydrologists have traditionally been able to ignore the kinetics of some major reactions. However, for predictions pertaining to radioactive waste disposal systems, involving tens of hundreds of years in systems where flow is slow, these reactions become important. Research is underway in the USGS to quantify them.

Characterizing the chemical and physical forms of plutonium, americium. and neptunium in contaminated natural waters is critical to the understanding of transport of these elements. In ongoing research in this area there is an attempt to establish the interaction of these elements with solutes in natural waters and to determine their behavior in a ground-water environment where migration is likely to occur. The key questions are: (1) What is the valance of these nuclides, and (2) Do they occur as a colloidal suspension?

Radioactive waste problems are dynamic, complex, and varied. Therefore, there can be no one simple solution to all problems. The different solutions will have to be imaginative, working primarily within the constraints dictated by the hydrogeologic environment at each proposed disposal site. The major factor will be to determine what types of radioactive wastes can be contained by the proposed site for the period of time required to isolate them from the biosphere

and hydrosphere. The form and handling of the waste should be tailored to the characteristics of the environment in the light of our understanding of the many natural processes that could affect waste containment.

DOE waste study: *N.M. demonstration for permanent site urged*—The Department of Energy March 15, 1978 issued a study that said spent fuel from nuclear plants, which would remain dangerously radioactive for over 100,000 years, could be safely stored in geologic formations.

At a press briefing March 15, John M. Deutch, energy research director for the Energy Department, said that the "earliest possible date" by which a national waste repository could begin operations would be 1988. Until recently it had been projected that a permanent disposal site would be ready by 1985.

The report urged that a test permanent disposal site be opened in New Mexico by 1983. The proposed New Mexico demonstration project, Waste Isolation Pilot Plant (WIPP) near Carlsbad, N.M., would be suitable for storing up to 1,000 used fuel assemblies in an area of no more than 20 acres. The spent fuel could be retrieved from the disposal site for up to 15 to 20 years if the government decided to approve reprocessing technology.

The study said that reprocessing of spent fuel was not a necessary part of a permanent waste disposal policy. There was not a significant difference between spent fuel that had not been reprocessed and that which had been reprocessed, according to the study. Some experts had suggested that reprocessing would ease disposal problems by reducing the volume of the wastes that had to be disposed. The Carter Administration, however, had opposed reprocessing, for the present at least, because it produced material that could be used for atomic weapons.

Deutch said that states would be able to veto plans to locate a permanent disposal site within the state. The precise way in which a state would exercise this right had not been determined.

Deutch said he did not know what action the federal government would take if all of the 36 states that were thought to have salt or hard rock suitable for repository sites refused to harbor the facilities.

The report said geological studies were being conducted on the known salt and hard rock formations throughout the U.S.

Cost estimates for the waste disposal program over the next 23 years ranged from $13 billion, assuming there was no growth in nuclear power, to $23 billion, if more nuclear power plants were opened. If the Energy Department charged utilities the full cost of disposing of their wastes, utility power costs would be increased by 4% to 5%, Deutch estimated.

The report said the federal government should take responsibility for all forms of nuclear waste disposal. In particular, the report recommended the government take over six commercial waste burial sites currently used for wastes of low radioactivity. "The longevity of management required," the report said, "clearly transcends most private enterprises."

The Nuclear Regulatory Commission should have licensing authority over all waste disposal operations, the report said.

The report drew criticism from some foes of nuclear power. Richard Pollock, director of Ralph Nader's Critical Mass Energy Project, said March 15, "We take major issue with the fundamental conclusion that radioactive waste can be disposed of safely in geologic formations." He continued, "It is somewhat dishonest to say this is a safe method when the site, technical design and material of the repository is unknown."

Senate Hearings on A-Waste Management

Senate hearings were held March 31, 1978 to discuss possible nuclear waste disposal programs being considered by the federal government. The hearings were held by the Subcommittee on Science, Technology & Space of the Senate Committee on Commerce, Science & Transportation. The hearings were opened following a statement made by Sen. Harrison H. Schmitt (R, N.M.). Excerpts from Schmitt's statement:

For the better part of 30 years, the Federal Government and the scientific community kept the issue of disposal of nuclear waste more or less in the closet. This closet was surrounded in the past by a basic distrust of the public's need to know and understand nuclear energy. Now that we need the public trust so that nuclear energy can fill an important, if not critical, role in our energy future, we find that we do not have that trust; at least not enough of it. Instead of building understanding among the public and politicians, we have built suspicion. In no area is this more true than in nuclear waste disposal.

I personally believe, as a geologist, that eventually it can be shown to be technically possible to dispose of nuclear waste in a variety of geological locations. However, I also personally believe, as a U.S. Senator, that geological disposal may be neither the most economical nor the most publicly acceptable method of disposal.

Then there is an urgent need not only for increased public and technical understanding of experimental salt bed projects like that near Carlsbad, N. Mex., but also of the alternatives to salt bed disposal. There may be an additional need for greatly increased funding so that these alternatives can be evaluated in a timely manner. ...

The need for a national policy on nuclear waste management is becoming increasingly critical. It is necessary to conduct full evaluations of the short-, mid-, and long-term implications of the various alternatives for the safe and acceptable disposal of nuclear waste, as well as the potential for nuclear waste utilization. This information will serve as an important part of the extensive evaluation that must be performed to facilitate policymaking processes....

For the past 30 years the management of nuclear waste has been the responsibility of the Federal Government. In the mid-1950s the Atomic Energy Commission, along with the National Academy of Sciences and the National Research Council, were asked to assess the nuclear waste disposal issue. At that time the study involved an evaluation of geological formations and the oceans. The National Academy of Sciences presented recommendations in 1957, which concluded that geologic disposal in salt mines was the most promising option, at least for the near-term. This type of burial was considered more attractive than ocean burial, primarily due to the lack of knowledge of the oceans with respect to all the consequences of such actions.

As a result of this decision, studies on nuclear waste disposal in salt mines were conducted throughout the 1960s and into the early 1970s. Investment in R. & D. during this time period averaged about $5 million per year. It was also during this time that the technology for solidification of nuclear wastes was developed which made the handling of wastes easier; the transportation of wastes safer, and long-term storage of wastes safer.

However, since this direction for handling nuclear wastes was established, many technologies have been developed and new scientific understanding has appeared which have added several possible options to the management of wastes.

First of all, our knowledge of the oceans has increased tremendously, and there have been significant advances in technologies that could be used for implementing this option. The possible international management of nuclear waste may require such an option.

Second, in the 1960s we saw the evolution of our space program, from its infancy in the early 1960s to a mature space program and, looking forward a bit, to the Space Shuttle. This experience in space has opened up the question of the feasibility of burying our wastes in space—another option only very recently considered.

Third, new technologies in chemistry and physics have raised the potential of utilization of nuclear wastes and/or specific isotopes contained in them. Examples of these technologies include laser iso-tope separation, photochemistry and conversion techniques.

Fourth, in addition to salt formations, there have also been some studies done on the use of granite, shale, limestone, and various volcanic rocks as potential nuclear storage formations.

Finally, new proposals include the potential of very deep geologic disposal by either in situ melting with layer isolation after a glass is formed, or by the implacement of stable crystalline phases con-taining the waste.

All of these advances, in addition to the technology of salt forma-tions, present interesting possibilities for disposing of or utilizing nuclear wastes in the short-, mid-, or long-term. It is time to bring discussions of all these alternatives to the public and realistically assess the potential of each one. We must think of not only what must be done for disposing of wastes in the next 10 years, but look to the next 20 years, and even 50 years, and assess what research and development must be done today to make the most attractive options possible. ...

The issue of nuclear waste disposal options is a controversial and often confusing subject. The existing technology for each option must be distinguished from future technology that needs to be developed. I don't think it is desirable to choose one option and just follow that path, but to open as many options as reasonable. We may exercise one option in the present, and others in the future as economics and technology change.

In a recent Harris poll, Americans were asked what factors will make this country a threat in the future. An overwhelming 91 per-cent place "scientific research" at the top of the list as a primary factor contributing to the success of the United States. 78 percent of the people surveyed believed "technological genius" is a principle factor. These figures suggest that even though we may not have the answer to nuclear waste disposal, the American people believe we can find that answer. ...

Roger W. A. Legassie of the Department of Energy's Office of Energy Research presented the following statement (abridged) at the hearings:

The formulation of a credible and broadly accepted nuclear waste manage-ment policy is a matter of the highest importance because we must ensure that nuclear wastes can be effectively isolated from the biosphere. Large quantities of nuclear waste exist today as a by-product of 30 years of production of nu-clear materials used in our National defense and from an increasing amount of electricity generation for domestic use. There is a legitimate public concern over potential disposition for these wastes. An effective and responsible nuclear waste management program that meets these public concerns is an important step towards assuring that commercial nuclear power will continue to play an important role in meeting our energy needs.

President Carter, in his April 29, 1977, National Energy Plan, directed that a review of the entire waste management program be undertaken. The Secretary and senior officials of the Department consider this review of the overall nuclear waste management program to be a matter of the highest priority. The Department has taken the first steps toward conducting that review and developing a comprehensive Administrative policy on nuclear waste management and a realistic program to implement this policy.

A task force was established by Under Secretary Myers to define the issues and options for dealing with nuclear waste. This task force was chaired by me with members from Energy Technology, Environment, Policy and Evaluation, Intergovernmental and Institutional Relations, Controller, General Counsel, and members of my staff. The report of that task force has now been released.

The Task Force report presents an assessment of the current nuclear waste management programs, identifies important issues, and explores alternative courses of action for resolving these issues. While the report contains significant recommendations, it is not intended to establish new policy or to commit the Department or the Federal Government to specific new programs or schedules. Rather, it hopefully should serve as the vehicle to stimulate discussion among a wider range of interested parties during the remainder of the policy formulation process. The issues raised by the task force will be addressed by the Administration after a thorough public review.

The Task Force report raises a number of issues with regard to present nuclear waste management policy and programs. I should like to highlight some of the very preliminary recommendations which I consider most significant for management of wastes from commercial power operations.

1. *A majority of independent technical experts have concluded that high-level waste (HLW) can be safely disposed in geological media, but validation of the specific technical choices will be an important element of the licensing process.*— Existing technical issues relate to the selection of the medium. Specific site and repository designs should be considered for each medium. Research is necessary to resolve medium selection and repository design issues. An accelerated effort to compile and analyze existing evidence bearing on geological disposal and feasible alternatives is also needed. The licensing process will assure that there is appropriate public scrutiny of our efforts and that the technical approaches taken are valid.

2. *Reprocessing is not required for the safe disposal of commercial spent fuel.*—From the point of view of the design at a repository for safe disposal, there is no significant difference between spent fuel and reprocessed high-level waste. Since a repository can be designed to accept either spent fuel or commercial HLW, disposal of spent fuel can be pursued initially.

3. *Consideration should be given to an early demonstration of the geologic disposal of a limited number of spent fuel assemblies in the Waste Isolation Pilot Plant (WIPP).*—This disposal should take place with full licensing. It should be compatible with the results of on-going R&D and it should employ conservative repository design characteristics.

4. *The Spent Fuel Policy announced by President Carter in October 1977 must be integrated with the Waste Management Policy.*—At issue here is the methodology that will be used to determine the one-time charge to utilities for the interim storage and subsequent disposal of spent fuel. Such a methodology must be developed and integrated with adoption of a detailed scenario for future storage/disposal. The principle of a "one-time" charge is essential, although utilities may be offered various storage/disposal options (bearing different costs).

5. *The Task Force report highlights the importance of an intermediate Away from Reactor (AFR) storage required between on-site storage of spent fuel at the utility reactors and ultimate disposal which will not be available before 1985.*—The character and amount of AFR storage required over time is sensitive to installed nuclear power, repository availability, and implementation of the spent fuel policy. Initial AFR storage is needed by 1983. Additional work is needed to define future interim storage capacity needs and to study ways for private industry implementation.

6. *The target for initial operation in 1985 of a National Waste Repository (NWR) for the permanent disposal of commercial HLW or spent fuel may not*

be met; this does not affect the mid 1980's schedule for WIPP.—The potential delay in NWR arises from the site selection process and a more realistic assessment of licensing requirements relative to previous plans. These considerations need not significantly impact the scheduling of WIPP as a near-term demonstration facility that site selections and evaluation are currently underway.

7. *The responsibility for the ultimate disposal for all forms of nuclear waste should rest with the Federal Government and long-term waste disposal facilities should be subject to NRC licensing.*—The importance of effective nuclear waste management and the national character of the production and disposal of waste may lead to an expanded Federal role. This issue should be carefully reviewed in light of its long-term financial and management consequences to the Federal Government. A licensing process that allows broad participation will lead to improved public confidence in long-term disposal.

8. *The NEPA process is an integral part of the nuclear waste management program and DOE efforts in this regard must be strengthened.* Additional effort is needed on the Generic Environmental Impact Statement (GEIS) on Commercially Generated Radioactive Waste because the GEIS will play a major role in the process leading to disposal of commercial waste.

9. *Policy and program management responsibility for Waste Management should be raised to a higher level in DOE.*—At present, the Director of the Waste Management Program reports to the Director of Nuclear Programs in Energy Technology as opposed to the Assistant Secretary. Because of the importance of nuclear waste management, the task force report recommends that program should report directly to the Assistant Secretary.

10. *There are substantial budgetary impacts of the Task Force recommendations and legislation would be required to carry out many of the suggested changes.*—Additional effort is needed on defining future budget inpacts under some of the recommendations and determining how these budget impacts should be financed....

Dr. Colin A. Heath, Assistant Director for Waste Isolation, Division of Waste Management, Department of Energy, told the subcommittee:

... As you mentioned, Senator, since the 1950's, numerous and varied techniques have been suggested for the disposal of radioactive wastes. And I would like to summarize what some of these options are, how they might be implemented, their current status, and why geologic disposal presently is considered to be the most suitable method in the immediate future.

When we look at the alternatives to our present approach of geologic disposal, there are really two basic types that we can consider. The first is variations of geologic disposal, which might entail additional processing of waste prior to the disposal in order to reduce the volume or change the character of the wastes as which might include the alternate geologic approaches which you mentioned, as well as the methods of rock melting, island or ocean bed disposal, and tectonic plate disposal. The second type covers basic alternatives to geology such as ice sheet and space disposal which change the nature of the barrier between the waste and man.

Perhaps I can just summarize some of the main thrusts of the current program first before going to the alternatives.

As you summarized very well, the initial recommendation to pursue salt beds made by the National Academy of Sciences in 1957, was really the beginning of the program. At that time, the members of that committee stated that the hazards associated with radioactive waste were such that no element of doubt should be allowed to exist regarding safety. They interpreted that at the time to mean that the waste should not come into contact with any living thing at any time after its disposal. .

In the context of that criterion, it was their recommendation that we should further pursue the option of looking a deep salt beds. That, at the time, was considered the best alternative. The only significant threat that they identified to the validity of this concept was the potential for the deep geologic repository being intercepted by flowing water, which might carry the radioactive constituents back to the biosphere.

However, they believed at that time through judicious selection of the site that the conditions of flowing water could be circumvented and thereby minimize the threat of the waste being reintroduced.

In addition, the potential for natural calamities such as earthquakes, volcanoes, or meteor strikes compromising the integrity of an underground repository were, and continue to be, considered. It is felt that if basic precautionary guidelines are followed, the worst potential impact of these calamities would represent only a minuscule threat to the respository's integrity. Subsequent reports by the National Academy have continued to support the technical feasibility of geologic disposal in bedded salt.

Other groups, including the American Physical Society and the United States Geological Survey, have stated, as you did this morning as a geologist, that acceptable repositories can be constructed, but have pointed out the additional R. & D. that must be performed to validate the technical choices concerning specific geologic media and sites. And, as Roger said, we certainly agree that there are specific technical issues about which more information is needed. That is why we are conducting a wide range of programs to obtain the required information.

If I can just briefly summarize our existing program. It's identified in two specific categories: The first is to identify potential sites and locations; and the second is the development of the technology for the development and the engineering support.

The geologic exploration studies are being conducted, as I said, to identify potential repository sites for commercial wastes. And since most investigations of geologic disposal have centered on salt formations, the primary emphasis on the program has indeed been on the known promising salt formations in the United States.

However, site suitability studies are also being undertaken at the DOE Hanford Reservation near Richland where basalt formations exist, and the DOE Nevada test site near Las Vegas, where I was yesterday, where there are shale, granite, and other nonsalt media which may potentially be suitable.

Then in the second part of the program, we do have extensive R. & D. which is continuing to provide the capability that we must have to analyze behavior mechanisms for radioactive waste placed in geologic formations. Included here are what we call a waste isolation safety assessment program, which is looking at the methodology to perform the assessment of long-term safety. Then we have studies underway to predict the structural behavior of repositories, borehole plugging studies to produce reliable exploratory hole and repository shaft plugs, a way to seal the repository after we are finished there, experimental in-situ tests to evaluate the heat-induced physical behavior of the media; thermal modeling, and waste media interaction studies.

And as you also mentioned, as part of the program we are also proceeding with the site evaluation and design of the proposed project near Carlsbad, the waste isolation pilot plant. The WIPP was originally conceived as a repository for transuranic-contaminated wastes generated in the national defense program, and to provide the capability to perform in situ experiments with high level wastes.

As you know, the waste management task force, which Roger referred to, has proposed that WIPP also be used for a moderate-scale demonstration of the capability for ultimate disposal of the spent fuel. However, the Secretary of Energy has not yet acted on this recommendation, and so it is not yet a departmental decision, and we are continuing to get more input into that decision....

Now, to the variations that I mentioned before—and trying to use the same structure of categories—first, looking at additional processing steps, we are conducting studies of what we call "partitioning" and "transmutation."

A partitioning would involve dividing the high-level waste into two parts: that which is short lived and high-heat generating; and that which is made up of long-lived, low-heat generating elements.

Transmutation would involve converting the long-lived radionuclides in high-level waste into either shorter lived or stabile isotopes, by neutron bombardment....

It should be noted that both of these techniques would require reprocessing as a necessary first step to separate the long-lived isotopes of concern. Both this and the subsequent partitioning would, of course, result in generation of additional wastes, including additional process chemicals, that would get mixed in with the radionuclides. So that this technique still does not remove the requirement for some kind of geologic disposal.

The transmutation process furthermore would require the use of fast neutron reactors, or special accelerators requiring the application of advanced technology which would have to be developed and put into place.

Another variation of this approach is the chemical resynthesis of high-level waste elements in order to reduce their mobility. In this scheme, these elements may be individually processed into synthetic crystalline minerals for disposal into similar natural geologic materials.

Several potential problems of scientific and technological nature have been identified. These include, once again, the increased amounts of contaminated materials from chemical processing; and, second, the requirement to perform this mineral synthesis in a radioactive environment. We feel that a major research program would be required to address these and other questions related to the feasibility of this option.

As an alternative to conventional geologic disposal, island disposal would involve the emplacement of wastes within deep stable geologic formations at depths reachable by conventional mining methods, but using an island as an entry point. The island would be used for the port facility, and the access tunnels, while at the same time providing a remote location, and possibly an international repository. This concept would be similar to land-based geologic disposal, except that the transportation would, of course, include over-water routes.

One issue of this option is the requirement to get consensus for the international repository. Another problem is the additional risk associated with overwater transportation of the wastes. There is no program presently being supported by DOE at this time on this concept because we feel that there are adequate formations using the same geologic disposal technique within the contiguous United States.

The basic concept of ocean bed disposal is to place wastes beneath the ocean floor in a geologically stable, biologically inactive region. Seventy percent of the Earth's surface is covered with water, with deepwater seabeds being among the least accessible of the formations on Earth. The potential, therefore, exists for identification of suitable geologic formations under the oceans. This concept is currently being investigated by DOE to determine its environmental and technical feasibility.

A three-phase program is envisioned: First, determine environmental feasibility; second, assess engineering feasibility; and, third, provide demonstrations of the concept.

Within the phase 1 period of 1978–83, the major objectives are to acquire oceangraphic, biological, and sediment data to establish that deep sediments in isolated regions of the ocean floor would be an effective barrier to the dispersal of radionuclides.

Provided that environmental feasibility is established, additional data would then be needed to allow further consideration of the option.

In addition to the U.S. investigations, an international effort is being conducted by the Seabed Working Group through the Nuclear Energy Agency of the Organization for Economic Cooperation and Development. Current international regulations and treaties prohibit dumping of nuclear wastes in the oceans. It is my understanding that studies of deep seabed emplacement are permitted, but that implementation would require modification to existing treaties.

Now, a variation of ocean bed disposal is tectonic plate disposal. This concept envisions emplacement in the deep sea trenches where the ocean tectonic plate is being driven beneath the continental plate. Ideally, the waste would be trapped within the sediments overlying the oceanic plate and could ultimately descend to a great depth within the Earth's mantle.

There are several major uncertainties about this. These trenches, unfortunately, are the most seismically and volcanically active zones of the Earth. The waters of trenches are characterized by high biological productivity and rich fishing zones. Also, the oceanic plate sediments have been observed to collect at the edge of the continental plate, rather than being pulled down underneath. So there is no assurance that the emplaced waste would indeed be deeply buried under the continental plate. There are no efforts currently underway to further investigate this particular concept.

The last alternate geologic approach, which you also mentioned, is the rock-melting disposal concept. This concept involves the emplacement of liquid or solid high-level waste produced by reprocessing of spent fuel into a small cavity in a particular geologic formation. If the thermal power density of the waste is high enough, and the emplacement geometry is appropriate, melting of the enclosing geologic medium could occur. The subsequent cooling of the molten rock to a glass or crystalline solid could provide a barrier to the migration of the radionuclides. No disposal technique involving melting has been extensively investigated to date, but the nature of this approach raises basic questions about feasibility and environmental acceptability which would have to be addressed.

A major research program looking at a very specific site would be required to evaluate this concept further.

Now, moving, then, to alternatives to geologic disposal, one alternative to geologic disposal which has been suggested is disposal in the ice sheet of Antarctica. Canisters of high-level waste would be allowed to freely melt through the ice sheet to bedrock below.

Other approaches would involve anchoring the canisters to surface markers for interim retrievability. Such disposal of radioactive waste in the Antarctic ice sheet is not currently considered to be a viable option. The American Physical Society report, for example, has strongly recommended against using this technique.

The other alternative to geologic disposal, of course, is the space disposal concept. This concept includes packaging waste materials in a safe manner and transporting them by rocket, or other means, to a location off the Earth. NASA is presently funding a program which is studying this concept in detail. And I understand that you have a NASA witness who will go into more details on that this morning.

My understanding is that the current NASA study is emphasizing the use of a solar orbit between the Earth and Venus for the emplacement of the waste with the possible use of a Moon crater for final disposal as an alternative. Both the solar orbit and Moon crater would provide long-term stability. The Moon crater option would require more energy consumption, but could provide a capability for retrievability.

DOE is cooperating with NASA in their evaluation by providing technical information concerning possible waste forms and quantities. One key fact, according to NASA, appears to be that the energy for space disposal of unreprocessed spent fuel elements would be a significant percentage of the energy originally made available from their use in power reactors.

It appears that some form of waste partitioning would be required for space disposal in order for it to be economically viable. It also appears that new waste forms would be required to allow survival of possible reentry conditions in the event of launch accidents.

Other issues are the possible requirement of partitioning of the waste to eliminate high specific radioactivity and heat-producing elements, and the need to design for maximum credible accidents in either launch orbit and/or reentry events which could impose significant extra waste form and packaging requirements.

While the space disposal option should obviously continue to be explored, it appears that development of necessary materials and processes to develop suitable waste forms could take considerable amounts of additional time and effort, and this technique cannot be really considered as a near-term alternative.

In summary, then, the alternatives to geologic disposal do not appear to be attractive for implementation in the immediate future due to the status of the technology and remaining problems and issues which we have identified.

There is a broad technical consensus that disposal in geologic formations is the most promising alternative at this time. A current program is structured to provide repositories for defense- and commercial-generated waste by the mid-1980's to late 1980's, although we recognize the validation of specific technical choices will be an important part of the licensing process that lies in front of us.

George D. De Buchananne, Chief of the Office of Radiohydrology, Water Resources Division, U.S. Geological Survey, provided the following statement (abridged):

The geological, biological, physical, and chemical aspects of nuclear waste containment have been of concern to earth scientists since 1955 when the National Academy of Sciences gathered 65 scientists in Princeton, New Jersey, to consider the many problems related to the disposal of these wastes.

Radioactive wastes can be referred to as high-level, transuranic, and low-level. For the purpose of this discussion, high-level wastes (HLW) include those wastes that have a high-level of penetrating radiation, high rates of heat generation, and a long toxic life.

In addition to those high-level wastes generated by the reprocessing of spent reactor fuel assemblies, one must also reckon with the spent fuel assemblies themselves if they are to be disposed of without reprocessing them for recovery of plutonium and uranium. Transuranic (Tru) wastes are described as those solid or solidified wastes that contain long-lived toxic manmade elements, generate little or no heat, and have radiation penetration levels of less than

1,000 millirems [1] per hour after packaging. Transuranic elements occur with high-level and low-level wastes. Low-level radioactive wastes (LLW) are those materials that have become contaminated through use and do not fit into either of the above categories.

Since the beginning of the Atomic Age, many disposal concepts have been considered in connection with radioactive waste management depending upon the classification of the particular waste being considered.

For high-level waste the various disposal concepts have included: (1) transmutation—the conversion of a radioactive nucleus to another isotope by bombarding it with radiation or nuclear particles; (2) extraterrestrial disposal—where waste would be placed in orbit around the sun: (3) ice sheet disposal—where the waste would be placed in the polar regions; (4) sea floor geologic disposal—where waste would be placed in the rocks underlying the ocean deeps, and (5) deep geologic disposal—where the waste would be placed in deep continental geological formations.

These five concepts of disposal have all been given serious consideration and research is still underway on several. After more than 30 years of study, the proposed disposal in deep continental geologic formations appears to have the best possibility for isolating the wastes from mankind, and the present discussion will be limited to this concept.

Until only recently (1970) low-level waste was permitted to contain transuranic waste. Although no final regulations have been issued, waste that includes more than 10 nanocuries per gram of transuranic waste is not being included in low-level waste but must be stored in a retrievable mode.

The waste management concept for transuranic waste since 1970 includes the packaging and storage of these wastes at an above-ground storage facility in Idaho awaiting the construction and licensing of a deep geologic repository. Intensive field and laboratory investigations of bedded salt deposits in New Mexico have been underway for several years to determine the feasibility of using these deposits as a host rock for a geologic repository.

The waste management concept for low-level solid radioactive waste has included two modes of disposal: (1) the packaging and subsequent disposal of low-level waste at sea, and (2) shallow land burial. Sea disposal has been discontinued so that now the management concept for all low-level wastes involves shallow land disposal. .

Within the past few years, in response to growing pressures for a resolution of the problem of disposing of wastes, earth scientists at various universities and government laboratories, as well as at the U.S. Geological Survey, began an intensive examination of the problem. As a result of this expanded examination, modified concepts of geologic disposal have evolved, and aspects of some older concepts are being recognized as requiring further study.

The basic requirement for safe disposal is that the waste should be isolated from mankind and the general biological environment for their toxic lifetime. The single most likely means of moving the waste is transport by circulating ground water.

High-level waste

The basic problem that must be addressed in all considerations of geologic disposal of HLW can be stated as follows. HLW material which is not in mechanical, thermal, or chemical equilibrium with its natural environment, is placed in a complex geologic host with the objective of not allowing physical and/or chemical migration of radionuclides from the disposal site into the biosphere in hazardous concentrations.

Therefore, it follows that the multicomponent physical-chemical system that will develop in and around the repository as a result of waste-rock interaction must be completely understood in order to avoid any unforeseen effects. The

---

[1] Rem: A unit of measure for the dose of ionizing radiation that gives the same biological effects as 1 roentgen of X-rays: 1 rem equals approximately 1 rad for X, gamma, or beta radiation.

effects of natural disturbance must be considered as well. The major categories of investigation needed to resolve these problems are: (1) investigation of natural and waste-induced processes; and (2) characterization of selected environments or regions for use in evaluating their potential for repositories.

The objective of these two categories of research is ultimately to assess all possible failure modes of waste repositories. Failure may result from both natural and waste-induced processes occurring independently or in conjunction with each other.

The Survey's ongoing high-level nuclear waste research program is conducted in cooperation with the Department of Energy (DOE). Early emphasis of the DOE's National Waste Terminal Storage program has been on disposal in salt formations with later studies of shale, crystalline rock, and other rock types. The U.S. Geological Survey has geochemical and hydrologic work underway for bedded salt formations and salt domes. Recently, a modest effort was started to evaluate the suitability of shale and crystalline rocks as host media by a combination of field and laboratory studies.

Investigation of natural and waste-induced processes includes:
(1) Geochemical interaction of the waste and its host media,
(2) Analysis of seismic risk and development of seismic design criteria, and
(3) Nature and effect of climatic changes.

Investigations directed toward the characterization of regions include:
(1) Evaluation of geologic remote-sensing techniques,
(2) Evaluation of regional and borehole geophysical methods,
(3) Hydrogeologic evaluation of salt domes,
(4) Hydrogeologic evaluation of bedded salt deposits,
(5) Hydrogeologic evaluation of shale, and
(6) Hydrogeologic evaluation of selected rock types of the Nevada Test Site.

Low-level waste

Six State-owned commercially licensed and five Federally controlled land disposal sites have been used in the United States. The 11 sites contain a total of more than 50 million cubic feet (1.5 million cubic meters) of solid radioactive waste that has been generated during the past 30 years. The Nuclear Regulatory Commission estimated in 1976 that by the year 2000 there will be an accumulation of more than 4.1 million cubic meters of low-level nontransuranic wastes.

After solid low-level radioactive waste is buried, the most likely method for its exposure, release, or movement from the burial site would be by a combination of erosional processes and transport in ground and surface water.

In the event that buried waste were to be exposed at the surface, the radionuclides could then be transported by surface water or by the wind as particulate matter. Present knowledge of the rates of geomorphic processes (involving erosion) and man's understanding of the probability of occurrence of earthquakes should be used to select sites so as to minimize the possibility of radioactive release through these processes.

Where solid radioactive wastes are placed beneath the land surface, they are subject to dissolution and transport by percolating ground water. Radioactive nuclides leached from the waste materials could then be transported downward and laterally as waste solutes to areas away from the burial site, or be transported upward by capillary flow to the overlying soil zone to be concentrated in plants growing at the land surface or concentrated in salts on the surface. In nature these transport processes are usually extremely slow, yet deliberate, and the subsurface flow paths of such solutes may be exceedingly tortuous.

In addition to the high-level waste program, the USGS–DOE cooperative program includes field investigations at low-level waste disposal facilities at two DOE disposal sites, one in the arid region of basaltic rocks of Idaho and the other in the humid environment of folded and fractured calcareous shales and limestones of east Tennessee.

Part of the low-level radioactive waste program, is directed at determining the processes and principles of radioactive waste migration. These investigations are both theoretical and applied. Theoretical mathematical simulation and laboratory studies related to radioactive waste disposal are already underway at USGS research centers in Reston, Virginia; Lakewood, Colorado; and Menlo Park, California. The Survey has also initiated comprehensive field investigations at the several state-owned low-level waste disposal sites, with the cooperation of the responsible state agencies in New York, South Carolina, Kentucky, Illinois, and Nevada.

UNRESOLVED PROBLEMS OF RADIOACTIVE WASTE DISPOSAL

Presently, geologic containment appears to be the most reasonable alternative for radioactive waste isolation if an acceptable geologic repository can be constructed. Because there will always be some residual uncertainty about the basic process of ground-water migration, waste-rock interaction, tectonism, and climatic variability, a conservative approach to the development of these repositories is called for. This approach should involve the development of clear and detailed statements of geologic criteria, the development of adequate techniques for verifying that a particular site meets these criteria, and a plan for long-term monitoring of the repository to ensure that the natural processes that actually occur are those that were anticipated or are at least benign.

High-level waste

One of the greatest gaps in current knowledge concerns the chemical interaction between the host media and the waste. A significant factor, for example, is related to the generation of heat by the wastes. The highest repository temperatures would occur during the first tens or hundreds of years. The release of thermal energy in the geologic media will result in a variety of mechanical, mineralogical, and hydrologic effects that may strongly influence transport of wastes away from a repository. These effects, some of which are imperfectly understood, are extremely difficult to predict for the intervals of geologic time required for the isolation of the wastes.

The problem of characterizing the environment of a potential repository is amplified by the necessity to minimize disruption of the integrity of the geologic containment media by excessive drilling of exploratory holes in the repository area. Geophysical techniques, particularly downhole techniques, need to be improved to adequately explore and describe the volumes of rock needed for a repository.

Most of the repository sites will be located within the stable part of the North American continent. However, because of the hazard of high-level waste, the risk of seismic activity exposing or releasing the wastes must be evaluated for every area considered for siting. Adequate methods for doing evaluations of this type that are meaningful for long time periods must be developed to avoid siting problems.

One of the major questions that will have to be asked about any proposed repository of radioactive waste will be the effect of future climate on the hydrologic regimen and geomorphic setting. Major climatic oscillations, with periods on the order of tens of thousands of years, have been a feature of global climate for at least the past million years and may be expected to continue. Therefore, existing paleo-climatological data need to be reviewed to judge the likelihood of the wastes being exposed during a future erosion cycle and/or transported as a result of a change in the hydrologic regimen.

In order for a repository to be acceptable, the host rock must be essentially impermeable. This raises the question of how to measure permeability in a drill hole in rocks of very low permeability. The technology for doing this needs to be developed.

Because circulating ground water is a potential mechanism for transporting wastes to the biosphere, the existing flow pattern must be completely known and models for contaminant transport must be developed. Mathematical models of fluid and solute transport in granular media are available, but flow in fractures, which is the principal avenue in crystalline rocks and shales, is not well understood. In general, further research is needed to develop methods of defining the spatial distribution of hydrologic parameters of rocks. In particular, the occurrence of fractures and the hydraulics of fracture flow should be investigated.

The assessment of containment properties of the rock at any potential high-level waste repository involves considering the potential movement of transuranium elements as well as other radioactive contaminants. There are several aspects of this assessment which need to be fully understood including the following: (1) chemical reactions between the radioactive contaminants and the natural earth materials through which they are being transported, and (2) the effect of the high-level waste heat on the migration of waste via ground water.

Drill holes, mine shafts, and other rock excavations in the repository are potential pathways that will have to be evaluated in terms of the role they might play in releasing wastes to the biosphere. All repostiory openings, regardless of depth, will have to be sealed by some, as yet unknown, technology which will in effect return the site, as nearly as possible, to its original condition. In addition, work is needed on the problem of monitoring a repository and the surrounding area for an indefinite time period to detect changes which might signal a loss of containment.

Low-level waste

The problem of designing engineered remedial action for the waste migration problem at the existing disposal sites is complicated by a lack of information on the detailed hydrology of the sea. The selection of future disposal sites, that will be required as the expansion of nuclear energy continues, will be dependent upon detailed knowledge of the hydrology of the proposed area plus a better understanding of several hydrologic factors that are directly related to the migration of waste.

In the solute transport of waste through a porous media, no model has been developed that will predict the concentration of a solute prior to actual introduction of waste into the aquifer. The theory is well described but actual field tests are lacking. Several studies are presently underway which will help resolve the obstacle.

As noted earlier, the flow of solutes through fractured rocks is not yet adequately understood. All verified transport models assume intergranular flow in the aquifer. However, where flow occurs in a fractured media these models are not applicable. Sound theory relating flow in fractured media to that in intergranular flow is presently lacking and must be developed before meaningful field tests can be made.

Theory for chemical reactions among water, waste, and earth materials in the unsaturated zone must be developed. Research on this subject is underway at the present time. After development of adequate theoretical concepts, the next step would be field testing and verification of the theory.

For simple chemical reactions of short duration under conditions of rapid flow, hydrologists have traditionally been able to ignore the kinetics of some major reactions. However, for predictions pertaining to radioactive waste disposal systems, involving tens of hundreds of years in systems where flow is slow, these reactions become important. Research is underway in the USGS to quantify them.

Characterizing the chemical and physical forms of plutonium, americium, and neptunium in contaminated natural waters is critical to the understanding of transport of these elements. In ongoing research in this area there is an attempt to establish the interaction of these elements with solutes in natural waters and to determine their behavior in a ground-water environment where migration is likely to occur. The key questions are: (1) What is the valance of these nuclides, and (2) do they occur as a true solute or as a colloidal suspension?

Radioactive waste problems are dynamic, complex, and varied. Therefore, there can be no one simple solution to all problems. The different solutions will have to be imaginative, working primarily within the constraints dictated by the hydrogeologic environment at each proposed disposal site. The major factor will be to determine what types of radioactive wastes can be contained by the proposed site for the period of time required to isolate them from the biosphere and hydrosphere. The form and handling of the waste should be tailored to the characteristics of the environment in the light of our understanding of the many natural processes that could affect waste containment.

John Healy of the Los Alamos Scientific Laboratory presented this statement (abridged):

Shallow earth burial has been used by man for years to isolate unwanted and noxious materials from his environment, often with little forethought as to potential risks of future uses of the land. At the time of the Manhattan District, sites were chosen for reasons other than waste disposal and it was natural to turn to shallow earth burial to dispose of the items that become contaminated with radioactive materials. This was done on the government controlled area, not only for convenience, but also to provide security against disclosure of the purpose of the project and some of its details. Following the war, this practice was continued with the burial grounds located on the site to avoid the possible spread of radioactivity to the public, such as could occur if the refuse were sent to the public disposal areas. This practice also provided an opportunity to study possible spread of the activity away from the burial site. During this period there was a wide-spread feeling that these burial areas were dedicated to this use by the government; that is, that there was a commitment by the government to retain these areas and not return them to the public. As a result, interests and concerns were largely devoted to the possible migration of radioactivity to ground water and rates of movement in these waters. This controlled the time required for the radioactive material to move to the boundary of the government controlled area and the concentrations that would be expected.

There have been changes in both attitudes and techniques in recent years that affect both the feasibility of disposal of certain materials by this method and the techniques used for those items that are buried. I would like to discuss some of these changes and their implications for shallow earth burial. My viewpoint will be that of one interested in protecting people from the ionizing radiations rather than that of a geologist, hydrologist, or the manager of a facility.

First, let us consider the types of materials that are typically sent to a burial ground. These, largely, consist of the miscellaneous items that could become contaminated during operations with radioactive materials and a few low activity process residues such as sludges from the treatment of liquid wastes or the ion exchange resins from the purification of reactor water. The mix of materials varies depending upon the operations at the facility generating the waste but can include paper, plastics, laboratory equipment, heavier equipment, such as lathes or milling machines, residues from treatment or diagnosis of patients in hospitals, and, even, the carcasses of experimental animals or plant materials resulting from research. The volume of wastes is increased by the practice of considering everything that enters an area where radioactive materials are handled to be contaminated. While this increases the quantity of material to be handled and buried, the sensitivity of instruments for measuring bulk wastes make this a prudent practice that is generally followed at all sites. Another source of periodic waste arises from the decontamination and decommissioning of facilities that have been used for radioactive materials. This provides equipment, building materials and the general trash that comes from the cleanup or demolition of buildings.

As an illustration of the magnitude of the radioactive waste problem, the National Academy of Sciences Panel on Land Burial has estimated that there were about 36,000 m³ (1,300,000 ft³) of solid wastes generated at DOE sites and about 57,000 m³ (2,000,000 ft³) at commercial sites in 1975. They predicted that the volume of DOE wastes would decrease because of increased treatment, but the volume of commercial wastes was predicted to increase rapidly because of the greater use of nuclear energy. This latter prediction will, undoubtedly, be affected by the President's policy on reprocessing. While these appear to be very large numbers, the Panel also noted that the DOE contribution was about equivalent to the solid waste from a typical U.S. town of 55,000 population. The overall U.S. generation of waste materials, excluding agriculture and mining—the two largest sources of solid waste—amount to about 15,000,000 m^3 (560,000,000 ft^3) per year, including about 340,000 m^3 (12,000,000 ft^3) per year of wastes that could be classified as hazardous.

Now that we have some information on the types and quantities of wastes that we are concerned with, let us turn to the changes that have and will affect shallow earth burial as an alternative for the disposal of radioactive wastes. An early indication of the changes was the decision by the Atomic Energy Commission in 1970 to limit the burial of wastes containing transuranium elements to those with a concentration less than 10 nCi/gram (one billionth of a curie per gram). Any wastes with concentrations above this value were to be stored in such a manner that the wastes could be retrieved, treated if necessary, and placed in a more secure repository. The 10 nCi/gram limit was based upon the levels of alpha emitting radioactivity found in some areas of the world. The presumption was, apparently, that if man had lived with these levels in his environment for all past time, it should not be inappropriate to increase the total of such areas by the addition of small areas at the same levels.

Another development in recent years has been the emphasis in DOE on improving the technology of shallow earth burial and reducing the volumes of waste. Studies of technology have included those on leaching and transport by water, uptake by plants on the surface of the buried material and studies of possible barriers to movement to be included in the burial grounds. These studies will be invaluable for future siting of burial areas and for the derivation of appropriate limits for the quantities and concentrations to be buried. Waste volume reduction has occurred by better control on the materials entering the work area and by some changes in processes. Research on the reduction in volume of the waste, once generated, includes studies of the feasibility of the use of devices to oxidize the combustible materials by methods such as incineration. Incineration has never been popular in this country because of the availability of cheap and convenient burial sites. It has, however, been used extensively in Europe where land is less available and, with the new emphasis in this country on the environment, there is little doubt that people will be willing to pay the price for this more expensive method of reducing volume and consequent land use. Such oxidation of the combustibles has other advantages including the elimination of possible decontaminating agents and other chemicals in the waste as well as the removal of organic materials that are subject to bacterial change. All of these items can result in mobilization of the radioactive materials with the resulting possibility of movement to ground waters. It will, however, increase costs of handling the wastes. An illustration of the improving technology is an interesting and potentially important development at the Battelle-Northwest Laboratory. This is the use of electropolishing, essentially removing a small layer of the surface by an electric current while in an electrolytic bath, to clean metallic items. This process is not yet in a commercial stage but is being investigated in several places. The process does produce a liquid waste with all of the radioactive materials removed from the metal and studies will be needed as to the best method of handling this waste. However, if the process can be applied commercially, it should greatly reduce much of the equipment and other metals going to burial, perhaps with some recovery and recycle.

A recent development has been the proposal of draft criteria for shallow earth burial by the Environmental Protection Agency. Possibly the most significant criterion is, "When institutional control is the method chosen to provide environmental protection of radioactive wastes, no restrictions on customary uses of associated land areas and surfaces and ground waters due to any residual risks should be required after 100 years." They also recommend that risks to future generations should not be greater than we allow to present generations as defined by our health and safety standards. These criteria are still under discussion and it is possible that there will be some changes, but a criterion of this nature will probably be accepted perhaps with some modifications based upon exploration of the practicality and feasibility. Such a criterion will, however, require limits based upon the concentrations and quantities of radionuclides that are buried.

At this point we should ask why any radioactive waste should be placed in shallow earth burial. The primary answer is the cost and the fact that, to date, it is the only alternative operable. The only proposed, and still to be developed, alternative at this time is the deep geologic waste repository. However, present planning indicates that the wastes must be treated to meet the repository criteria (for example, by incineration to eliminate the combustibles), packaged to meet the Department of Transportation regulations, shipped to the repository, and emplaced. The exact increase in cost will depend upon the type of waste and degree of treatment needed, the distance of transport, and the actual design of the repository. However, a rough estimate would indicate the possibility of an increase of about an order of magnitude or so.

In summary, the shallow earth burial of radioactive wastes will continue to be a viable alternative for the disposal of radioactive wastes of certain types. However, it is believed that, in the future, upper limits for the burial of a variety of radionuclides will be applied so that very long term care of the site will not be needed and additional criteria and engineering will be applied to the site so that the shorter term problems of migration will be controlled.

Dr. H. D. Sivinski of the Applied Biology Isotopy Utilization Division of Sandia Laboratories in Albuquerque, N.M., and Livermore, Calif. told the subcommittee:

The fission products and transuranic wastes of the nuclear fuel cycle contain large amounts of potentially valuable metals, industrial radiation sources and specialized energy sources. Sandia Laboratories has a major role in assisting the Department of Energy in developing the technology and investigating the potential of using some of these products in processes which historically have used materials and energy sources now considered to be in short supply. If these products, now, considered to be a national liability, were to be separated and utilized, they could well become a needed national asset.

The Sandia sludge irradiation program is part of a larger DOE effort to develop beneficial uses of individual isotopes. It is jointly funded by the DOE and EPA through an interagency agreement and addresses the potential of using gamma radiation from a fission product, cesium-137, to reduce the pathogen levels of municipal sewage sludges sufficiently to permit their use in unlimited agricultural applications. Over five million dry tons of municipal sludge are being produced each year in the U.S. and with the passage of PL 92–500, the Water Pollution Control Acts Amendments of 1972, the requirement for secondary treatment will probably double this amount by 1985. Cessation of ocean dumping is mandated to end in 1981 and other disposal options by cities are being rapidly foreclosed by stricter environmental controls, suburban development near treatment plants and landfills, public pressure and the uncertain availability of fuels for energy intensive disposal options such as hauling and/or barging to remote areas or incineration. PL 94–580, the Resource Recovery and Conservation Act of 1976, suggests land application of sludge to facilitate resource recovery and conservation of this potentially significant national agricultural resource. Such application would also contribute to the solution of the rapidly escalating sludge disposal problems faced by most metropolitan centers.

Two major problems are involved however in the agricultural use of domestic sludges. Heavy metal contaminants are taken up by plants and animals in the food chain, and control over the heavy metal pollution sources must be enforced. The second problem is the health hazard due to the pathogens (bacteria, viruses and parasite ova) in sludge. Biological inactivation of these pathogens with gamma irradiation from waste cesium-137 is the major Sandia research and development activity in this program. Current research results indicate that a 500 kilorad dose to composted or dried raw sludges will probably be sufficient to reduce the pathogens to desired levels for general agricultural use. (The current standard for radiation sterilization of medical supplies and disposables is 2,500 kilorads). Two agribusiness uses are being researched by New Mexico State University (NMSU) using Albuquerque and Las Cruces, NM sludges. The classical use, land application for both fertilizer value and soil conditioning, is being done in greenhouse and field plots. Significant improvements in plant yields over optimal chemical fertilizer formulations are found when raw irradiated sludge is applied at equivalent nitrogen rates. Ruminant animal feed supplementation during dormant range periods is also being investigated. Sludges are rich in energy and nitrogen and can be used in lieu of the soybean or cottonseed meals normally used to balance a range animal's ration during the 2–3 month supplemental feeding program required annually.

Values for sludges used in these programs have been determined to be between $5–60/dry ton as fertilizer and up to $100/ton as a feed supplement. The radiation treatment cost is $18/ton. In contrast, disposal costs average $250/ton for trenching-in the sludge from the Blue Plains plant in the Washington, D.C. area and range nationally from "give away" to $400/ton. Before we can realize the potential of irradiated sludge, the process of radiation disinfection needs to be demonstrated on a meaningful scale. ...

The isotope is diverted to beneficial use for two half lives or 60 years, and then reintroduced into the waste disposal program. The 75 percent reduction in activity in this time interval contributes positively in decreased isotope disposal requirements.

The Sandia Irradiator for Dried Sewage Solids (SIDSS) is a pilot plant for radiation treatment of composted or dried raw sludges. It is shown schematically in Exhibit B and will treat 16 tons of material per day to a dose of 500 kilorads. It is now under construction and will be operational in the fall of this year. It will use 1 megacurie of cesium-137 from military wastes being encapsulated at the Waste Encapsulation and Storage Facility (WESF).

A larger 25 ton/day EPA-funded demonstration plant is planned for the Washington, D.C. area and will use 3 megacuries of WESF capsules. This facility will assist in the transfer of this technology to the public sector.

Successful technology transfer depends greatly on the economics of the process. A cost/benefit study of this program was done by Sandia, Battelle, and A. D. Little, Inc. The Federal Register price of $0.10/Ci for the WESF capsule was used in the analyses and it was shown that with a 10 percent penetration of the 10 million dry tons/yr market, annual savings will approach $100 million. It should be emphasized that this annual economic benefit is possible now without reprocessing commercial power reactor fuels. Sufficient amounts of cesium-137 for treating 10 percent of the national sludge are available from reprocessing of military wastes. Approximately 200 MCi are at Hanford and a comparable amount is expected from Savannah River over the next 12 years.

The question of increased isotope availability and cost from commercial fuel in order to meet greater needs in this and other programs such as fruit and vegetable disinfestation was addressed in a thorough study recently completed by Exxon Nuclear Inc. Separation flow sheets were evaluated for various isotopes and a cost of $0.08 to 0.13/Ci for cesium-137 was obtained when an isotope separation facility was designed to operate in conjunction with a commercial reprocessing plant. This study validated the basic assumption of the cost/benefit study which had shown a favorable ratio using the $0.10/Ci price.

The use of another isotope, technetium-99, for marine biofouling and corrosion inhibition is also being investigated. Biological experiments leading to understanding of the mechanisms involved are being done in conjunction with the University of New Mexico. When completed, the information gained from this work could contribute positively to solution of both fresh water and marine cooling water problems in electrical power generation stations and to programs such as the DOE Ocean Thermal Energy Conversion (OTEC) program.

In summary the Sandia program is attacking the isotope and sludge utilization problem on many fronts. We believe we are pursuing the research areas required for a successful sludge utilization program which must be cost-effective, resource and energy conservative, socially acceptable and technically reasonable. Such results would provide another option to cities burdened by the sludge management problem. A cost reduction in waste treatment programs is possible while recycling a valuable resource for food production.

Dr. G. Ross Heath of the Graduate School of Oceanography at the University of Rhode Island told the subcommittee:

I will try to summarize the concepts of seabed disposal that we have heard mentioned here this morning, and to review recent advances in the area. I should emphasize right from the start that this is a geologic concept; it differs from land options only in the sense that the seabed is under 3 to 4 miles of water. This may be an engineering problem, but in some ways it is a geologic advantage!

The seabed program has been running at full speed for about three years with a budget accelerating from less than 900,000 in 1976 to about 3.1 million in the current fiscal year. The program is being carried out jointly by people from Sandia Laboratories, who are providing management and much of the expertise in the engineering and thermal areas and scientists from a number of universities, primarily major oceanographic institutions around the country, who are bringing to bear their expertise in problems unique to the oceans.

The basic problem we face is to determine whether there are areas in the deep ocean basins which are stable from geologic, oceanographic and climatic points of view, and which are predictable over time scales of millions of years. This viewgraph shows the outlines of the continents and also the position of recent earthquakes. You can see that the earthquakes are concentrated around the edges of large so-called crustal or lithospheric plates. In terms of geologic stability, then, we are concentrating our attention on the centers of these plates; the areas where there are no earthquakes. In order to have low biologic activity, we find that we have to focus on the central North and South Pacific or the central North and South Atlantic areas where there are great oceanic gyres characterized by very little biologic productivity.

An additional criterion is the absence of any mineral resources. Consequently, we have to stay away from thick sedimentary sections which could contain recoverable hydrocarbons, and also from the area where manganese modules will be mined in the next few years. This area lies east and south of Hawaii. Incidentally, recent work on manganese modules suggests that the ones rich in copper and nickel are very closely associated with high biologic activity in the surface waters. Thus, in a sense, our second and third criteria are not independent.

To sum up, we feel that the most attractive areas are the mid-plate, mid-gyre areas in the North Pacific and North Atlantic. Alternative areas have been suggested by other workers. We have already heard mention of trenches. We feel that they are not attractive because of their high seismicity (you can see all the earthquakes centered along the coast of South America beside one of the major trenches). In addition, a lot of the surficial material carried into trenches seems to get scraped up against the continents rather than carried down into the mantle.

Another suggested disposal area is fracture zones in mid-ocean ridges. Again, we do not favor such sites because sedimentation is very unpredictable due to slumping of material into the fractures, and because of the possibility of erosion by bottom currents which often channel through fracture zones.

Similarly, we do not favor burial in deltas, where sedimentation is very rapid, again because slumping has been recorded in these areas.

We are attacking the Seabed Disposal problem in a systems engineering sense. We are trying to pin down the natural processes that might affect buried waste, as well as the technical questions that will be involved in preparing waste material for disposal. In addition, of course, there are political, economic, and social questions which we have looked at in only a rudimentary way thus far. We feel that we have to establish environmental and technical feasibility before we can justifiably delve too deeply into some of the other aspects.

In looking at the seabed, we have focused on a multiple barrier concept. We view buried waste as being separated from man by a series of barriers. The obvious first two are the waste material itself and the enclosing canister. Variations in the properties of these materials can be engineered, but probably these two barriers will be effective only for hundreds to perhaps thousands of years. As far as natural barriers are concerned, the first is the water column. We dismissed this one very early because of physical mixing and extraction and transport of dissolved wastes by organisms. The ocean likely would provide a barrier for at most a few hundred years.

Similarly, the second possible natural barrier, the rocks of the oceanic crust beneath the sediments has to be downgraded because work by the Deep See Drilling Project has shown that such material is highly fractured and has permeabilities ten thousand times as great as the overlying sediments, even when it is quite old. By elimination, then, our efforts are really focused on the final natural barrier, the clay sediments laid down in the centers of mid-plate, mid-gyre areas.

Pelagic clay sediments are attractive from many viewpoints. They are highly impermeable; it would take a dissolved nuclide something like a million years to diffuse a hundred meters, even assuming no interaction with the sediment. There is absolutely no natural flow in these sediments, a feature which distinguishes them from virtually all continental deposits. Work on samples collected by the *Glomar Challenger*, for example, has shown that profiles of dissolved elements in the pore waters are exactly as predicted from pure diffusion. There is no evidence of any flow. Our initial work on the sorption properties of pelagic clays suggest that a thousand to a hundred thousand times as much of the dissolved material is attached to the sediment as is in solution in the pore water. As a result, diffusion is a thousand to a hundred thousand times slower than the rate I just mentioned. Clearly, once we get to diffusion times of a billion years for a hundred meters, we have an essentially infinite barrier.

At the same time as evaluating these natural barriers, we have begun to look at the effects of perturbing the environment, calculations of the results of inserting a heat source, for example were surprising, at least to me. If we were to bury a canister of high level waste or used fuel in deep-sea clay, we would find that during the thermal lifetime of the waste material, a water particle that started against the top of the canister would migrate only a little over a foot, again reflecting the extreme impermeability of those sediments.

Everything is not yet rosy, of course. A couple of important problems are still unresolved. The first of these stems from the fact that putting a heat source into the sediments will lead to a marked decrease in the density of the pore water. The water will never boil (we are talking about pressures of hundreds of atmospheres), but for a 600° centigrade temperature rise, for example, its density will decrease by about a factor of ten. We are concerned that this decrease may result in sinking of the canister, or rising of the entire mass of sediment carrying the canister with it. Obviously a lot more work has to be done on this problem.

Another area of concern is the way in which waste material could be emplaced in the deep seabed. Some initial calculations and simple laboratory tests suggest that very rapid emplacement may be the best way to go. Under these conditions, the sediment heals immediately behind the canister, eliminating the problem of having to backfill the hole. The sediment seems to regain its mechanical, engineering, and chemical properties very rapidly. Again, though, these are very preliminary results and a lot more work has still to be done.

In terms of the development of the program, we are presently in Phase One—the Environmental Assessment Phase. We expect to establish whether or not the concept is environmentally feasible within the next two to four years, thereby determining whether or not to go on with the program. Obviously, if one or both of the two outstanding problems I just mentioned turns out to be unsolvable, we will have to simply give up on the program.

Assuming this does not occur, Phase Two will require an additional five years for assessment of the engineering feasibility of the concept. Questions of emplacement then will become very important. Do we use a free-fall type of emplacement?, or do we have to use a drill ship or other more complicated emplacement technique? If Phase Two is successful, Phase Three, which presently is targeted to be completed in the early 1990's, will involve actual demonstrations of emplacement.

An aspect of this program which looms large compared to land disposal options is its international implications The deep seabed is now, and the areas we are looking at are likely to remain, in international waters. As a result, it is unlikely that the U.S. could, or perhaps should, view sealed disposal as a solution that can be adopted unilaterally. A number of other countries are interested in the concept, and an International Seabed Working Group has been set up under the auspices of the NEA Radioactive Waste Management Committee.

We feel that if burial is environmentally feasible and can be carried out at very little additional cost relative to dumping, we should do everything in our power to encourage other nations to consider the option because of the enormous increase in safety that will result.

Perhaps I can sum up by saying that the seabed option has a number of advantages. The areas we are interested in are extremely remote from any of man's present or likely future activities. In addition, and perhaps more importantly, Seabed Disposal may provide a potential international solution to the whole question of radioactive waste disposal, which could have important repercussions on the nuclear proliferation question.

At the same time, there are still a number of potential disadvantages: monitoring and retrieval techniques have not yet been demonstrated; emplacement techniques have not yet been demonstrated; special port and vessel facilities that are unnecessary for disposal within the continental United States will be required; and, finally, although the question is somewhat controversial, it appears that Seabed Disposal may be contrary to existing international laws and agreements, although such an interpretation depends very heavily on whether one views this as dumping rather than geologic disposal.

At this stage, we are still cautious. We feel that our role still is one of assessment, rather than advocacy. Nevertheless, our studies to date have not identified any technical or environmental reasons to suggest that high level waste or spent fuel could not successfully be emplaced in stable sedimentary formations beneath the abyssal floor of the deep ocean....

Uranium Milling

The processing of uranium for use in nuclear plants begins with the milling of the uranium-containing ore. Currently there are 19 uranium milling operations in the U.S. Nine of these mills are licensed by the Nuclear Regulatory Commission (NRC), and the remaining 10 are under state control.

Uranium milling results in the accumulation of sandy radioactive waste—tailings—that, unless safely disposed of, constitute environmental and health hazards. These U.S. mills now hold 110 million tons of tailings, and by the year 2000, the NRC estimates, they will have nearly one billion tons of uranium waste.

Over the past 30 years, 22 uranium mills in the Western states and Pennsylvania have become inactive. They left behind, with some 25 million tons of unprotected tailings, a health hazard and a gigantic cleanup cost estimated at up to $200 million. A large part of the cleanup cost is due to the fact that during the 1960s much of the uranium waste had been used in the construction of homes and buildings.

President Richard M. Nixon June 16, 1972 signed a bill authorizing $5 million as aid to Colorado where uranium tailings had been used in the construction of homes. Detectable radiation levels were found in 5,300 homes in the community of Grand Junction; in some cases the levels exceeded health standards. The AEC was authorized to defray up to 75% of the costs of efforts to reduce the radiation exposure there.

Colorado Gov. John A. Love had said in 1971 that since the federal government was the sole uranium customer, it had sole responsibility to pay the estimated $12 million–$20 million removal cost. The AEC had refused to take legal responsibility for plants built before the 1969 Environmental Policy Act, but it said that waste control would be required before any new mills were licensed.

As part of the continuing attempt to find an effective legislative program to reduce radiation exposure from these tailings and to take remedial action at the abandoned mills, congressional hearings on three bills were held June 26-27, 1978 by the Subcommittee on Energy & the Environment of the House Committee on Interior & Insular Affairs. Rep. Morris K. Udall (D, Ariz.), chairman of the subcommittee, said in an opening statement (abridged):

The three bills before us—H.R. 12535, H.R. 13049, and H.R. 12938—have been submitted by myself, for the Department; by Mr. Evans of Colorado, and by Mr. Marriott of Utah, respectively. Each bill offers a slightly different approach for a proposed cleanup of uranium mill tailings.

The mill tailings covered by these measures were created in the production of uranium under contract to the Federal Government for our nuclear defense programs. At the time the contracts were in effect, adequate regulatory control over the tailings hazard did not exist, and tons of the sandy material accumulated in ponds and piles without protection from leaching, wind, and erosion. Some of the tailings have been used in construction of homes and buildings. There, the concentrated effect of the radioactive gas emitted by the material has created such a threat to public health that foundations and walls have had to be removed at great expense to the State of Colorado and the Federal Government.

In my opinion, and I think this is reflected in the Department of Energy's approach, the Federal Government now has a responsibility to assume the burden of cleaning up these sites. I have several concerns, however, regarding the implementation of such a program.

First, I believe that the States should share some of the costs as well as the benefits of the program. But I want to offer them some flexibility and prevent their costs from being so great as to preclude their participation.

I want to insure that all agencies involved—DOE, the Nuclear Regulatory Commission and the Environmental Protection Agency—have all the proper authorities they need to accomplish their tasks.

I want to insure that when remedial actions are undertaken at the sites, the actions are adequate to protect the public health and the environment for the necessary period of time. ...

*Rep. Dan Marriott, (D, Utah) testified
June 26 (abridged):*

Two weeks ago the State of Utah's Bureau of Radiation and Occupational Health announced two very important findings: One, that they had developed, in cooperation with EPA's Radiation Laboratory at Las Vegas, a new technique which significantly reduces the time needed to measure the level of radiation; and two, that by using this new measurement, a 1.5 working level of radiation was detected at Firehouse No. 1, which was built on tailings from the Vitro Chemical site in Salt Lake City. They detected the highest reading in the sleeping quarters of the firehouse. Dr. Gibbons, director, Salt Lake County Health Department, and Dr. Olson, director, State of Utah Health Department, are here to testify this morning and can give us the details.

It is my understanding that this 1.5 working level measurement is one of the highest ever taken. This means that the hazard these tailings are creating is at least as great, and possibly greater, than that existing anywhere in Grand Junction, Colo., which has generally been thought to have posed the very highest danger. It was that high degree of exposure that prompted Congress to take action in 1972 to address the most immediate concern in Grand Junction. By way of comparison, the radon gas exposure at Firehouse No. 1 is seven times that found in an average uranium mine....

I feel very strongly that the Federal Government must bear the responsibility to pay for the cleanup of these sites. After all, this problem is due to the past neglect of the Federal Government.

Beginning in the late 1940's, the Federal Government, as part of a national defense program, encouraged the production of uranium ore, was the sole purchaser, established the purchase price, chose the contractors, approved the processing sites, and provided the necessary technology. In short, they have exclusive control of this program.

In fact, let me quote briefly from the Atomic Energy Act of 1954, subsection 161 I(3), which authorized the AEC to:

> I. Prescribe such regulation or orders as it may deem necessary * * * (3) to govern any activity authorized pursuant to this Act including standards and restrictions governing the design, location, and operational facilities used in the conduct of such activity in order to protect health and to minimize danger to life or property.

From this it is very clear that the Congress, in authorizing the Atomic Energy Commission to undertake this program, intended that they should have exclusive jurisdiction in this area.

At the time the AEC entered into this program, there was a significant lack of knowledge regarding the hazards of uranium processing. In fact the AEC recognized that they were acting in an area that posed a lot of unknowns.

Very early in the program, January 1950, the AEC prepared a document entitled "Waste Materials in the United States Atomic Energy Program," by Abel Wolman and Arthur E. Gorman, pointing out this problem. Let me quote from that document:

> By congressional act the Atomic Energy Commission has virtually sole power over this important development. In this respect it is unlike any other industry with which the public has hitherto been confronted. Normally the activities of an industry, which potentially may have some public health influence upon the surrounding populations, has the continued scrutiny of official public health agencies on Federal, State, and local levels. In this case of nuclear fission operations, however * * * the normal process of supervision and regulation generally available through Federal, State, and local health departments and health officials are missing, and their initiation is handicapped by security policies.

Other levels of government particularly State and local, were not consulted. This was, in the very purest form, strictly a Federal program.

Now, many years later, we have the advanced technology to clearly establish the serious health hazard that these uranium tailings have created, yet the very legislative proposal submitted by the Department of Energy, H.R. 12535, calls for the States to pay one-fourth of the cost of Uncle Sam's shortcomings....

That means that those States who bear the total burden of the environmental risks will be required, in addition, to come up with a full 25 percent of the total cleanup costs. Using the States of Utah and Colorado as examples, they will have to contribute roughly between $8 to $10 million each. Yet, they had no input into establishing any of the rules or regulations which governed this program....

Above all, we in Congress want to avoid a situation where we establish a mechanism for the completion of the remedial action and then never see it undertaken because of the inability of the States to raise their 25 percent of the total cost.

This funding issue is a critical consideration to the people of all the affected States because of the immediate need to take action with respect to these tailings piles. May I point out that the Salt Lake Vitro site is on a 128-acre tract located only 30 city blocks southwest of our State capitol building. The tailings pile is largely uncovered and subject to continuing wind and water erosion. The most alarming thing is that the site is only partially fenced and is easily accessible to the public.

While the tailings cleanup should be a Federal responsibility and clearly a Federal program, I feel very strongly that the States need to be actively involved. They are responsible for protecting the health, safety, and welfare of their citizens. They must determine what action is necessary to remove this continued risk. By giving the State the responsibility for the development of the site restoration plan, you assure protection over the long and short terms to the public. However, I don't for a minute envision each State operating within a vacuum. Any remedial action will have to have the concurrence of the Secretary of Energy....

Rep. Austin J. Murphy, (D, Pa.) testified June 26 (abridged):

I am here on behalf of a group of citizens from my home district to provide you with their concern over the existence of radon gas emanating from uranium mill tailings....

In presenting the testimony, I urge all of you to carefully consider the human aspects of the problem before you. The radiological and engineering surveys conducted by both Federal and State agencies are voluminous and detailed. However, there has been too little said about the impact that the discovery of the radiation is having on the lives of the people involved. Indications from the reports are that there is no great short-term hazard to their personal health. However, many of the people from the Canonsburg area have worked on these sites for as long as 10 years. The potential for serious health problems will undoubtedly increase as the result of that exposure and must be accurately determined.

In addition to the personal safety of the people of the area, there has already been damage to their local economy and to their personal economic security. Industrial tenants have been lost and cannot be replaced. Employment has been reduced due to the loss, and those industries who have positions open for workers cannot fill them so long as the radiation threat exists, and expansion programs already started have, of course, been suspended....

Once their needs are met, we can then seek the scientific and engineering solutions that will remedy the remaining problem.

... And I would like to call to your attention, and it is attached to my testimony, a letter dated February 1966 from the Atomic Energy Commission.

In summary, what the AEC at that time said, and I quote:

We have concluded that due to the insignificance of the contamination which may be present in your former Canonsburg facility, no hazard to health and safety is involved.

Based upon that statement, the local industrial development concern in my home county took this site from Vitro, developed it, created and sold land to various industries, hired people in the local area, and consequently, it was the Federal Government that actually enticed people to use this site, not only as Mr. Marriott and the Chair have pointed out, that there were lack of Federal regulations at the time the uranium tailings were disposed of. They then went one step farther and encouraged the redevelopment of this site and the location of people actually working upon it, so I think that not only should we consider Mr. Marriott's legislation of which I am a cosponsor, but we absolutely must go one step farther, we must consider the social and economic problems that have actually been created by the Federal Government, and then get to the scientific elimination, but I first say that the people there and their health and their economic well being must be considered.

> Dr. Lyman Olsen, Director, Utah
> State Division of Health testified June 26
> (abridged):

It is my opinion that a final solution of the health hazard resulting from the tailings pile located in Salt Lake Valley must be made, and that the solution is to move the pile to a remote location where it can be stabilized.

This tailings pile is located in a residential-industrial area, and is adjacent to a much larger population center than any other pile in the country. Thousands of people work and live in close proximity to the pile and are exposed to radioactive dust, radon gas, decay products of radon, gas and gamma radiation. This is not the only pile of this type in the country, but it most certainly constitutes the greatest threat of any because of its location in the center of a rapidly growing urban-industrial complex of over one-half million population.

From 1951 to 1964, the Vitro Minerals and Chemical Co. processed 1.7 million tons of uranium ore on its 128-acre site in South Salt Lake. In 1965, the mill was converted to vanadium production and was in operation until 1968, milling 106,000 tons of vanadium-bearing material. The total amount of tailings deposited at this site is estimated at 2.3 million tons, excluding the contaminated subsoil.

The tailings, which were under the regulatory control of the Atomic Energy Commission, now Nuclear Regulatory Commission, have never been under control of the State of Utah.

On September 1, 1959, the AEC authorized Vitro to dispose of the tailings by selling them to Talboe and Harlin Construction Co., and/or Salt Lake Suburban Sewer Districts. Tailings were removed and used in various construction sites throughout the Salt Lake Valley, but this practice was stopped when the Utah State Division of Health learned about it and objected.

In June 1962, the State Division of Health started limited investigations of the tailings pile, and verified that there were significant amounts of radium in the tailings.

In June 1967, a letter was received by the Division of Health from Vitro, "strongly suggesting that this material should be employed for fill where large quantities are used in comparatively small areas, such as high construction." The Division of Health objected but encountered strong pressure in favor of the proposed use. The idea was later dropped.

Also in June 1967, the U.S. Public Health Service and the U.S. Atomic Energy Commission, in cooperation with the State health agencies of Colorado and Utah, started a joint project to evaluate the emanation of radon-222 near uranium tailings piles. A report of this study was published in November of 1968 and clearly identifies the health hazard associated with radon and radon daughters near uranium tailings piles.

As pressures increased for development of the land near the tailings, the Division of Health adopted a policy that it would not support any development within one-half mile of the tailings pile. This was based on the realization that health questions would arise repeatedly if occupied buildings were constructed in any area where radiation build-up could occur as a result of emanations of radon from the tailings. This policy is still in effect.

By July 1969, it was apparent that the use of uranium tailings material for construction in Grand Junction, Colo., did in fact represent a health hazard to the occupants of homes and businesses.

Further studies led to development of guidelines for remedial action. These guidelines, as published by the Surgeon General of the United States, required removal of the tailings or other remedial action where gamma levels were greater than 0.1mR/hr1 (1mr/hr, milliroentgen per hour, a measure of gamma radiation) or where radon daughter concentrations exceeded 0.05 WL2 (2WL, working level, a measure of alpha radiation from short-lived radon daughter elements).

In 1972, EPA conducted a gamma survey in Salt Lake County and located 71 anomalies which by close screening resulted in the location of some 17 sites to which tailings had been hauled. These included five residences, seven businesses, and five lots. Unfortunately, EPA was unable to complete this survey due to a lack of funding; obviously, this survey must be completed prior to or as part of action to solve the Vitro hazard.

On September 18, 1973, the State Division of Health requested that the AEC assist the State in the study of the feasibility of removal of the tailings pile. This resulted in the Energy Research and Development Administration contract with Ford, Bacon & Davis, Utah, Inc., to conduct a phase I assessment of the Vitro pile and a subsequent phase II architectural-engineering report. This report was published in April 1977 and substantially agrees with the previous studies and offers 10 options for remedial action.

A review of the report by the Utah Council of Science and Technology led to endorsement of two options, both calling for recovery of tailings from other sites in the valley followed by complete removal of the pile to a remote area.

The importance of early action was emphasized in recent weeks when our staff discovered that certain occupied buildings in Salt Lake County to which tailings were hauled, presumably during construction, show internal levels of radiation far exceeding currently accepted limits. The more detailed procedure necessary to accomplish these measurements was made possible by development, by a physicist on our staff, of a new procedure for developing radon levels.

In these instances, radon and decay products resulted in readings up to 1.5 WL, which is 30 times the upper limit for remedial action prescribed by the Surgeon General, as previously noted, and 150 times the level at which no remedial action would be considered. Readings were later confirmed by the EPA Radiation Laboratory at Las Vegas. People occupying these buildings clearly need better protection than they are getting, and removal of the tailings from under and around the structures is the only alternative to abandonment of the buildings.

The gamma radiation on and around the present tailings site range from approximately 0.05 to as high as 3 mrem/hr. The radiation levels in front of the occupied dwellings directly across the street (south side of 3300 South) were as high as 0.2 mrem/hr. In the yard of one of the neighboring production facilities (Wondoor Corp.), levels varied from 0.2 to 0.5 mrem/hr and the level was 2 m43m/hr approximately 50 feet north of this building.

In the waste water treatment plant yard north of the pile, the radiation level was as high as 0.15 to 0.3 mrem/hr. Gamma spectrometry of the tailings and water from the nearby ponds indicate that the bulk of the radiation was from radium-226 and its decay products. One of the decay products of radium-226 is radon-222. This gas is particularly hazardous since it can be inhaled. Upon inhalation, the alpha particles emitted by the decay of radon are known to be particularly injurious to lung tissues.

The permissible radiation level for occupationally exposed personnel, for example, badged workers at an NRC facility, is 0.1 rem for a 40-hour workweek. For the general populace, this level should be reduced by at least a factor of 10. For young people, the recommended maximum permissible dose is significantly less than this.

At a meeting in 1969, attended by staff members of the Division of Health and representatives of EPA and AEC, at the Division of Health offices at 4th South and State Street in Salt Lake City, the

EPA representatives verbally reported that levels as high as 0.3 to 0.5 WL had been measured in some of the businesses adjacent to the tailings site and expressed the conviction that no further monitoring was necessary to establish existence of a hazard.

Since the milling of uranium ore began in 1949, the knowledge about the decay products of radium has increased markedly. On August 7, 1959, the AEC stated that trace quantities of thorium and 0.029 percent uranium posed no threat and we encouraged use of the tailings as fill.

As recently as February 11, 1966, the Director of Regulations for AEC wrote the Joint Committee on Atomic Energy: "From a radiological safety standpoint, licensing control of such tailings is not required."

In .1969, in response to a direct question, Mr. Paul Smith from the EPA office in Grand Junction, said:

I feel no further monitoring is necessary to substantiate that the tailings do constitute a clear health hazard and that the radon levels exceed existing health standards.

Three, other representatives of AEC and EPA concurred. Almost 10 years later, and after a number of additional studies, the results are not only verified but the risk was probably underestimated.

In June of last year, the Regional Director of EPA stated:

In the predominant daytime wind direction, there was a significant power curve relationship of radon concentrations to a distance out to a mile and a half from the center of the tailings pile.

Add to this the continued exposure for an additional 10 years, the increase in population over that period of time, the regulations on exposure of pregnant women, the risks of children with their activity and increased metabolic rates, and you have a serious problem. Every new guideline that has come out has lowered the amount of acceptable exposure. The National Academy of Sciences stated: "No exposure to ionizing radiation should be permitted without the expectation of a commensurate benefit."...

Total relocation of the radioactive material seems to be the only logical solution. Although many suggestions for stabilization in place have been made in past years, they have been discarded as ineffective or impractical. It is significant to us, and a continual worry, that each time new and better scientific information becomes available—as in the case of our new technique for measuring radon and its daughters—the extent of hazard is concluded to be worse than previously thought. ....

Under these circumstances, it is folly to speculate that new knowledge will provide a scientific breakthrough which gives us the techniques to contain radon and other radioactive elements at the tailings site. Utah residents have already suffered more than reasonable exposure and want the pile removed. ...

Dr. Anthony Robbins, Executive Director, Colorado Department of Health testified June 26 (abridged):

As you know, during the 1950's, the Federal Government and the mill in Grand Junction encouraged people to take the waste tailings, which were described by everyone as being safe at the time, and these were put in, used as fill under thousands of structures in Grand Junction.

We have now completed the survey of about 25,000 structures, and the remedial action project is now almost complete. But one of the things that we are only now getting around to, and it is kind of unfortunate, is to do some of the health studies that were never done in the past.

The first thing that we did, not quite 2 years ago, was review the cancer statistics in the State, looking at the types of cancer that are known from the other studies to be related to radioactive exposure for men. In doing so, we found in Mesa County, where Grand Junction is located, apparently elevated rates of lung cancer and acute leukemias.

We have since searched the cancer registry in Grand Junction and the registry for the rest of the State and are no longer as concerned, because when age adjustment is done, the apparent increase in lung cancer mostly disappears.

We are still left at this time with a small number of cases but apparently twofold increase in the rate of acute leukemias for people living in Mesa County. This cannot at this point, and should not at this point, be attributed to the radioactivity, though that is a possible explanation, because the next part of the study is yet to be done. ...

Decommissioning reactors. Nuclear power reactors as well as other fuel cycle facilities reach the end of their useful lives in about 40 years as a result of either obsolescence or adverse economic problems. Scientists and government officials, therefore, have started to worry about the problems of decommissioning and decontaminating nuclear reactors. The NRC estimates that at least 12 commercial nuclear reactors face decommissioning action by the year 2000. There are three primary decommissioning plans used by the NRC. Two of them, mothballing and entombment in place require radiation monitoring and protective security systems. The third in dismantling, the removal of all radioactive components.

Rep. Leo J. Ryan (D, Calif.) July 25, 1978 discussed the problems of decommissioning obsolete nuclear reactors.

Ryan quoted from an article, published July 5 in the San Francisco Chronicle, about Andre Cregut, who built most of the major nuclear powerplants in France and who heads the French program for dealing with obsolete nuclear plants. According to the Chronicle, Cregut said "that nobody has been able to dismantle a commercial atomic reactor, ... [and] [w]ith dozens of nuclear plants reaching obsolescence throughout the world, scientists and governments have begun facing up to the troubling problems of ridding the landscape of these dangerously radioactive structures, and estimates for the clean-up operations are running into the billions of dollars.... Already 20 nuclear power plants have been closed in the Western industrialized world—15 in the [U.S.] ... and five in Western Europe.... By the year 2000, there will be more than 100 inactive atomic plants."

Cregut concluded that "[e]ven if we entombed these plants, there is no way to be certain that after 500 or 600 years the protective casing will be physically maintained or guarded. . . . Do we have the moral right to leave these plants in place knowing that it will take hundreds, perhaps thousands of years before they cease to be dangerously radioactive?"

International Developments

Nuclear waste storage. The Atomic Energy of Canada Ltd. revealed Sept. 27, 1972 that the agency planned to keep spent radioactive fuels from nuclear reactors above ground in "engineered storage facilities."

The wastes, in solid cylinders, were to be tied together to form fuel bundles and stacked so that natural air circulation could be used to keep them cool. Piles of fuel bundles were to be stored in concrete structures the size of an enclosed football stadium.

Australia: *plutonium waste issue*—The disclosure that a pound (0.5 kilograms) of plutonium in a "discrete mass" was buried at Maralinga in South Australia prompted opposition leader William Hayden to ask for a full inquiry Oct. 9, 1978 into the disposal of radioactive waste from British nuclear weapons tests in Australia in the 1950s.

Hayden said a study was needed to determine whether the plutonium at Maralinga could be obtained by terrorists for use in a weapon. Also, Hayden said, the inquiry should find out whether the storage methods at Maralinga complied with international safeguards and whether any of the radioactive waste buried there came from outside Australia.

It had been revealed in the press that in addition to the pound mass of potentially recoverable plutonium, about 43 more pounds (19.5 kilos) of plutonium were buried in Maralinga in unrecoverable form.

The disclosure had embarrassed the government, because on Oct. 5 Ernest Titterton, the head of the safety committee for the Maralinga tests, had firmly denied that any discrete mass of plutonium was buried at Maralinga. Within hours of Titterton's denial, Defense Minister James Killen confirmed that such a mass did exist.

Acting Foreign Minister Ian Sinclair announced Oct. 10 that Australia had accepted a British offer of a team of technical experts to "help Australian authorities define the extent of possible physical problems involved with the Maralinga site."

Sinclair's announcement was interpreted by Australian observers as indicating that Britain had refused to comply with the wish of the Australian government that Britain remove the radioactive wastes left over from the nuclear tests.

A statement issued by the British High Commission in Canberra Oct. 10 denied that British authorities had been guilty of any carelessness or negligence in cleaning up the after effects of the weapons tests.

International Safety Issues

Non-Proliferation Safeguard Controversy

A major aspect of the nuclear safety issue involves diversion of nuclear fuels or wastes for weapons use. Controversy arose repeatedly over national and international safeguards designed to prevent such misuse of nuclear materials.

3-nation accord provides safeguards. Britain, the Netherlands and West Germany agreed March 11, 1969 to build two plants to produce enriched uranium by a new gas centrifuge method. The agreement which provided safeguards against, the military application of nuclear material, called for the establishment of two organizations, one to operate the two plants to be built in Britain and the Netherlands, and the other to supervise manufacture of enrichment plants in all three countries.

The three nations signed a treaty March 4, 1970 in Holland for cooperative development of producing enriched urnium. The process could lead to a relatively inexpensive means of separating, according to density by centrifugal force, fissionable uranium isotope 235 from the nonfissionable uranium isotope 238.

The treaty, signed by Dutch Foreign Minister Joseph M. A. H. Luns, West German Foreign Minister Walter Scheel, and Britain Minister of Technology Anthony Wedgwood Benn, provided for the construction, commissioning and operation of plants at Almelo and at Capenhurst in Cheshire, England.

Spokesmen for the three powers emphasized that the treaty provided adequate safeguards against military application of the centrifuge process and that it in no way contravened the rules of Euratom, by which the Dutch and German governments were bound. (Scientists had warned that centrifuge separation could become an easily-hidden process for production of weapons grade fission materials.)

Woldzimierz Natrof, a Polish delegate attending the U.N. Conference of the Committee on Disarmament declared Feb. 24 that the three nations producing enriched uranium were proceeding too quickly with their investigations and that possible military applications of the process would imperil the nuclear non-proliferation treaty unless controlled by the U.N. International Atomic Energy Agency.

A-safeguards pact drafted. The U.N. International Atomic Energy Agency

(IAEA) announced March 11, 1971 that its safeguards committee, responsible for insuring enforcement of the Nuclear Non-Proliferation Treaty (NPT), had concluded work on a draft agreement to be negotiated with the treaty's signatory powers.

In recommending a compromise settlement of the dispute about payment of the cost of safeguarding the treaty the committee advised that payments continue to be met from the IAEA's ordinary budget but that assessments be changed to ease the burden on countries with low percapita incomes.

(The IAEA said that Finland March 19 and Austria May 4 had "completed negotiation of and initialed the text of an agreement relating to the application of safeguards by the agency, pursuant to Article III.1 of the Treaty on the Non-Proliferation of Nuclear Weapons." No details of the agreement were given.)

Dr. Rudolf Rometsch, the IAEA's inspector general, reported March 24 that the agency had started recruiting 55 inspectors to begin work in March 1972 on verifying compliance with the non-proliferation treaty.

Nuclear pact safeguards accords—Finland June 11 became the first nation to formally sign an agreement with the IAEA providing for application of IAEA safeguards to all Finnish nuclear material to insure their peaceful use. The agreement was in fullfillment of Article III of the Treaty on the Non-Proliferation of Nuclear Weapons.

Austria signed a safeguards agreement with the IAEA Sept. 21.

Uruguay Sept. 8 and Poland Sept. 16 initialed the text of safeguards agreements following the conclusion of negotiations with the IAEA.

The agreements were based on guidelines established by the IAEA nuclear safeguards committee in March and approved by the board of governors April 20. The guidelines included provision for a national control system by each state, coupled with the IAEA's right to independently verify the national findings.

The IAEA board of governors, meeting in Mexico City Sept. 22, 1972 approved a cooperation agreement with the five non-nuclear members of the European Atomic Energy Community (Euratom). It provided for coordination of IAEA and Euratom safeguards systems to verify that the atomic material was used peacefully. Under the accord, Euratom would conduct the basic inspection but would adapt some of its procedures to conform with IAEA requirements. The Euratom Council of Ministers had already approved the agreement.

The accord would cover the current members of Euratom, except for France, and two of the future members, Denmark and Ireland. France and Britain were not included because they already possessed nuclear weapons.

The board approved a safeguards pact with Mexico Sept. 22.

The IAEA held its annual general conference in Mexico City Sept. 26–Oct. 3, 1972.

The conference approved a cooperation agreement between the IAEA and the Agency for the Prohibition of Nuclear Weapons in Latin America (OPANAL) and approved pacts with three nations applying IAEA safeguards to the nations' nuclear material in accordance with the nuclear non-proliferation treaty, it was reported Oct. 3. The three nations were Mauritius, Morocco, and the Netherlands, the latter acting on behalf of the Netherlands Antilles.

Church outlines U.S. A-safeguards policy. Sen. Frank Church (D, Ida.), at Massachusetts Institute of Technology May 2, 1977, outlined the history of the U.S. safeguards system to prevent the spread of nuclear weapons. Church said:

Almost 25 years ago, the United States admitted that we could not prevent other countries from acquiring and developing nuclear technology. In his "Atoms-for-Peace" address to the United Nations on December 8, 1953, President Eisenhower outlined a new American policy based upon the recognition of this truth. He proposed the creation of a new organization under the UN to assist countries in developing nuclear energy for peaceful purposes. Recipients of this assistance were to agree to international inspection, in order to prevent diversion of materials for weapons use. As the world leader in nuclear technology, the United

States sought to influence the peaceful development of this vast new energy source while, at the same time, restricting its use in building weapons arsenals.

So we took the lead in the creation of the International Atomic Energy Agency (IAEA). Ever since, the IAEA, with our strong support, has furnished technical assistance and materials to participating countries and has developed and administered a system of safeguards. Meanwhile, the United States, carrying out its commitments under the IAEA Statute and the Treaty on the Non-proliferation of Nuclear Weapons, has helped other countries acquire, research and power reactors. We have been the world's major supplier of nuclear materials, equipment, technology and training. But the infant system of international inspection and control that we fathered has not kept pace with the growing need.

Church went on to introduce an international system of supply and control of nuclear technology. He said:

I would, therefore, suggest that we abandon any new and demonstrably futile attempt to contain the flow of nuclear technology and try instead to channel it in directions that will provide adequate supplies of energy while minimizing the potential for making nuclear weapons. We should aim toward eventual international ownership and control of the fuel cycle necessary to sustain commercial reactors in all non-weapons states. Regional uranium-enrichment and reprocessing plants should be established under the auspices of the IAEA. This agency should own all the nuclear fuel in its global system, be responsible for the fuel's security, and accountable for its use. Fuel elements could be leased to the participating nations on the condition that the spent rods would be returned, and that each reactor in which they are inserted would be subject to IAEA inspection.

No plutonium or other weapons material would ever be produced in a form which could be used for weapons fabrication; the fuel would be blended with other materials to make it unworkable for bomb construction. For example, the reprocessing would produce only blended uranium and plutonium, usable for fuel elements, but unworkable in a weapon.

Fuel fabricating facilities would be built immediately adjacent to the chemical separation plant, as would a facility for glassification of all their waste for deep burial. New fuel elements could be slightly irradiated as a last step before being shipped from the facility. Thus, they would require the same heavy shielding that is necessary to return irradiated fuel elements from a nuclear power plant.

Such a system of international supply and control would not only provide maximum protection against diverting fuel to weapons but would meet the legitimate demands of the non-weapons states for an assured supply of nuclear fuel at a fair price. Just as the American Government now sells enriched uranium to commercial reactors in this country, the IAEA could make its fuel available to participating states at cost.

The system I have described would also furnish the best insurance against terrorists. By dealing only in irradiated fuel, it would be necessary to steal at least one, and perhaps several, 50-ton shipping casks, in order to obtain enough material to make a single weapon. Furthermore, it would be necessary to extract the weapons material from the hot fuel rods, a feat requiring elaborate facilities. This should reduce the potential for illicit diversion for weapons material virtually to zero.

I do not underestimate the formidable task of establishing such a global system. But while we work in that direction, our national policies should constitute the example of how such a system might work. In this regard, I agree with Congressman Mike McCormack, the Chairman of the House Science and Technology Subcommittee, who recently observed:

"Of course no system is perfect or foolproof. It should be obvious, however, that this nation and the world can have the energy required for economic stability and a high degree of nuclear security as well.

"If some nation decides that, in spite of all the attendant problems, it is determined to obtain nuclear weapons, then it will obviously be much easier to make the weapons outside the fuel cycle, as was done by India. This could be done secretly today, regardless of any restraints the United States places on its own energy programs. It is more likely to happen if the United States does not provide leadership for a workable program to assure adequate controls and supplies of energy.

"Above all else, this nation must lead, and from a credible position. The other nations of the world do not believe that we can provide them with nuclear fuel unless we have a breeder program and unless we recycle and reprocess our fuel."

U.S. links cooperation to 'security arrangements.' The U.S. AEC disclosed its readiness to begin "exploratory discussions" with 10 nations on construction of gaseous diffusion plants used in production of fissionable uranium-235 for nuclear reactors, the Washington Post reported July 30, 1971. The decision was based on the working out of security safeguards that would keep the nuclear fuel from being used in nuclear weapons.

The nations involved were Britain, Japan, Canada, Australia, France, West Germany, Italy, Belgium, Luxembourg and the Netherlands. The U.S. had previously opposed sharing the formula for the separation of uranium-235 from ordinary uranium because it was a key element in the production of nuclear arms.

An agreement to help build the plants and share the formula would depend on the conclusion of "appropriate financial and security arrangements."

At the time the AEC enriched uranium for the power reactors of France, Belgium, Netherlands, Luxembourg, Italy, West Germany, Switzerland, Spain, Sweden and Japan.

U.S. seeks private uranium processing. In a major policy turnabout, AEC Chairman James R. Schlesinger called on private industry Dec. 8, 1972 to build the uranium enrichment facilities needed to supply the U.S.'s electric power needs. The AEC published proposed regulations for corporate access to secret government technology involving the enrichment process, until now a carefully guarded government monopoly because of its use in the production of atomic bombs.

In a related move, Schlesinger announced the same day that the commission would draw up new, more businesslike contracts for the agency's sales of enriched uranium to the electric utilities using nuclear power reactors. The new contracts would be more specific about the quantity and delivery dates of the needed uranium and would carry a penalty for cancellation.

Schlesinger said private industry would also be needed to maintain the U.S.'s lead in the export of nuclear power technology. He predicted that export sales of U.S. enriched uranium and reactors would total $3.5 billion annually by the mid-1980's, compared with the present $900 million.

U.S. to send more A-fuel abroad. A bill amending the Atomic Energy Act was passed by the Senate July 11 and House Aug. 1 and signed by President Ford Aug.

17, 1974. The legislation retained safeguards on use of the fuel. The major change permitted the AEC to increase the amount of nuclear material it distributed to groups of nations for peaceful purposes. The amount could be increased above statutory ceilings if both houses of Congress did not disapprove the increase within 60 days.

The provision retained Congressional control over such distribution. The AEC had sought elimination of the requirement for statutory approval of the distribution levels.

The bill also permitted export of small amounts of nuclear material for peaceful purposes to countries not having a nuclear-cooperation pact with the U.S.

Curbs set for nuclear sales. In an attempt to prevent nuclear weapons proliferation, major exporters of nuclear materials, except for France and China, had compiled a list of equipment they would supply to other nations only under assurances it would not be diverted for explosive use, the New York Times reported Sept. 24, 1974. The decision was set forth in governmental letters to the IAEA in Vienna.

The list detailed what the supplier countries considered to be special fissionable material and equipment for using it.

Among the supplier countries agreeing to the new curbs were the U.S., Great Britain and the Soviet Union.

U.S. warns against A-spread. Fred C. Ikle, director of the U.S. Arms Control and Disarmament Agency, warned at a news conference April 9, 1975 that "several countries, not now nuclear-weapons states, appear to be making determined efforts to acquire a capability that would enable them to build their own atomic bombs." He said that "the spread of nuclear weapons capability is riding on the wave of peaceful uses of the atom," that the transfer of peaceful nuclear technology was providing "not only the means, but also the cover" for a spread of nuclear weapons.

Sen. Edward M. Kennedy (D, Mass.), a speaker at the conference, criticized the major nuclear superpowers, the U.S. and the Soviet Union, for not living up to their 1968 treaty obligations to stop testing weapons and reduce stocks and for not keeping sight "of the ease with which other nations can build nuclear weapons."

(Ikle did not specify the countries appearing to push for nuclear status, but the New York Times account reported learning "authoritatively" that they included Argentina, Brazil, Israel, Libya, Taiwan, South Korea and Pakistan.)

Continued arms threat seen. The Stockholm International Peace Research Institute said June 12, 1975 that the global arms budget for 1974 totaled $210 billion despite reductions by the Warsaw Pact and the North Atlantic Treaty Organization (NATO).

Institute director Frank Barnaby warned in a press conference releasing the Institute's 1975 yearbook, World Armaments and Disarmament, that the major problem was continued stockpiling of growing numbers of nuclear weapons. He stated that the danger of conflict was heightened by the "catastrophically large trade in major weapons [to] third world countries, especially those in the Middle East." He said such trade had increased by 40% between 1973 and 1974.

Barnaby also described as a "dangerous situation" the promotion by U. S. Defense Secretary James R. Schlesinger of the large-scale deployment of "mini-nukes" in Western Europe to replace more conventional A-weapons. He further drew attention to the "not exactly marvelous" NATO storage facilities in Europe housing 7,500 nuclear weapons and asserted that if the "facts were more widely known . . . it would create concern among Western European countries."

U.S. gives 'mini-nuke' pledge—The chief U.S. delegate to the U.N. Conference of the Committee on Disarmament, Joseph A. Martin Jr., had presented a statement to the Geneva forum May 23, 1974 in which the U.S. gave assurances that it would not develop a new generation of miniaturized nuclear weapons, known as "mini-nukes," which could be used interchangeably with conventional weapons on a battlefield.

Critics of these relatively small atomic arms had argued that such weapons could make it easier to decide to cross the threshold from conventional to atomic warfare; the U.S. in its statement pledged that it would not act to "erode the distinction" between nuclear and non-nuclear arms. Reiterating, Fred Ikle said that the effect of the policy was to "establish a barrier against taking advantage of available technology to go into new types of nuclear weapons that make sense only if viewed as substitutes for conventional arms."

(According to U.S. sources quoted in the New York Times May 24, one purpose of the policy statement was to prevent the Atomic Energy Commission from pushing ahead with development of such a new family of weapons. Congress had denied a 1973 Pentagon request for $1 billion to begin producing new 8-inch and 155-mm atomic artillery that broached the mini-nuke category, it was reported May 24. However, the Pentagon was reportedly seeking funds this year to develop a much smaller nuclear warhead, described by the military as more effective against military targets with less likelihood of causing extensive damage to surrounding civilian areas. This, according to U.S. officials quoted in the Washington Post May 24, would, if approved by Congress, qualify under the new policy statement. Ruled out would be a return to the Davy Crockett missile [which was retired in 1967] and such items as "atomic hand-grenades" and tiny nuclear land mines.)

Lilienthal recommends export ban— David E. Lilienthal, the first chairman of the AEC, called Jan. 19, 1976 for an embargo on U.S. exports of nuclear reactors, materials and technology to halt the "terrifying" spread of atomic bomb capabilities. He made the recommendation at a hearing of a Senate Government Operations Committee panel headed by Sen. John H. Glenn Jr. (D, Ohio).

Lilienthal, calling the U.S. "the atomic patsy of the world," said many citizens would be "shocked" if they realized the extent to which the U.S. "has been putting into the hands of our own commercial interests and of foreign countries quantities of bomb material."

His call for an embargo was endorsed by Nobel prize-winning physicist Hans Bethe, who worked on the U.S. atomic bomb project during World War II, and physicist Herbert York, who worked on the U.S. hydrogen bomb project later, but they said the embargo should be a temporary one until an accord were reached on security for nuclear trade.

Canadian-Pakistani talks snag on waste issue. Canadian External Affairs Minister Allan Maceachen said March 2, 1976 that Canada and Pakistan had suspended weeklong negotiations on the continued sale of uranium fuel for a Canadian nuclear reactor in Karachi. President Zulfikar Ali Bhutto had talked about the matter with Premier Pierre Elliott Trudeau in Ottawa Feb. 23–25 and the discussions were taken up by other Pakistani aides after Bhutto left Ottawa to continue his Canadian tour.

The negotiations were snagged over Canadian fears that a nuclear reprocessing plant purchased from France earlier in the week could transform the wastes from the Canadian reactor in Karachi into material from which an atomic bomb could be manufactured. Canada insisted on imposing stricter safeguards for its facility, but Bhutto objected to Canadian jurisdiction over a non-Canadian plant.

A Canadian official had said Feb. 25 that "We have deep suspicions of anyone" who wanted a nuclear processing plant after Ottawa's experience with India.

Bhutto had said Feb. 25 that the United Nations International Atomic Energy Agency had approved the purchase of the French reprocessing plant Feb. 24.

Objections to the French-Pakistani deal had been expressed Feb. 23 by Fred Ikle. Ikle told the U.S. Senate Foreign Relations Committee that "the reason for the Pakistani interest in a reprocessing plant is the Indian development of nuclear explosives."

Nuclear export rules tightened—The government Dec. 22 put into effect greater restrictions on the sale of nuclear-reactor fuels and technology to other countries.

Announcing the new policy, External Affairs Minister Donald Jamieson said: "Shipment to non-nuclear weapon states under future contracts will be restricted to those which ratify the nuclear non-proliferation treaty or otherwise accept international safeguards on their entire nuclear program." Previous Canadian restrictions had applied only to items supplied by Canada, leaving receiving countries free to develop weapons with materials imported from other sources.

The new regulations placed Canadian exports under "the most stringent conditions in the world," Jamieson said, adding: "For all practical purposes, nuclear cooperation with Pakistan is effectively at an end." Pakistan had rejected placing its nuclear reactor, purchased from Canada, under the new restrictions.

U.S. & Pakistan discuss French A-plant— U.S. Secretary of State Henry Kissinger said Aug. 9, 1976 that the U.S. and Pakistan had agreed to settle their dispute over the projected sale of a French nuclear-processing plant to Pakistan. Kissinger had discussed the matter with Prime Minister Zulfikar Ali Bhutto during talks in Lahore Aug. 8–9.

Without providing specifics, Kissinger said Bhutto had consented to work out a compromise formula that would provide assurances by Pakistan that it would not divert nuclear material from the processing plant into atomic explosives. Kissinger suggested that this could be accomplished by a Pakistani-French agreement that would give Paris a veto over any Pakistani diversion plans.

The U.S. would bar the sale of about 100 A-7 Corsair jet-fighter bombers to Pakistan unless it agreed to a compromise on the A-plant dispute, Kissinger said.

The French Foreign Ministry Aug. 9 called in U.S. Charge d'Affaires Sam Gannon to express displeasure at what it regarded as American efforts to block the sale of the uranium processing plant to Pakistan. Gannon was told that France intended to complete the transaction. Ministry officials disclosed that the sale had been approved by an accord signed March 18 by France, Pakistan and the

International Atomic Energy Agency. The agreement gave the IAEA supervision and control over the plant to make certain that it would be used only for peaceful purposes. The U.S. member of the IAEA's Board of Governors had endorsed the pact, a ministry spokesman noted.

French A-export guidelines. French Foreign Minister Jean Sauvagnargues told the National Assembly (lower house of parliament) April 8, 1976 that the nation's export of nuclear material and technology would be "subject to controls to ensure that they are used for peaceful purposes."

Sauvagnargues said that the government would demand that "the bulk of our exports of nuclear material and equipment" be placed under the supervision of the United Nations International Atomic Energy Agency in the importing countries. These countries would be required to commit themselves to using the supplies exclusively for peaceful purposes, he said. The French government, Sauvagnargues said, would demand that importing countries rigorously safeguard the transport and use of nuclear material.

In another development, President Valery Giscard d'Estaing told visiting Prime Minister Robert Muldoon of New Zealand April 16 that France would hold no atmospheric tests of nuclear weapons in the Pacific. Giscard d'Estaing did not rule out underground tests in the area.

French-West German nuclear pact OK'd—
Representatives of the French and West German governments signed an agreement in Bonn May 18 on the exchange of information and technology on the development of fast-breeder nuclear reactors. France had an acknowledged lead in the research and development of such reactors which, when operational, were expected to yield both electrical power and additional nuclear fuel.

Under the terms of the accord, the two governments agreed to spend between $84.7 million and $105.9 million annually on reactor development. A French-West German consortium was established to supervise the pooling of information and technology and the sale of production

licenses to other nations. The consortium partners were the French Atomic Energy Commission; Navatome, a French builder of atomic reactors that was 40%-owned by Creusot-Loire, and Interatom, a subsidiary of the West German company, Kraftwerk Union AG. The U.S.-based Westinghouse Electric Corp. also participated. It held exclusive licenses for certain nuclear processes involved.

The signing of the pact followed the announcement April 15 in Paris that the French government had given the go-ahead for the construction of a 1,200-megawatt prototype fast-breeder reactor, called "the Super-Phoenix." The reactor, which would cost an estimated F4.5 billion (about $940 million), would be built at a site north of Lyons. Electricite de France, the state-owned utility, had a 51% share in the project. Other participants included French and Italian companies.

French-A-plant purchase canceled—
South Korean plans to purchase a nuclear processing plant from France were canceled, apparently because of strong U.S. pressure, it was confirmed by U.S. and South Korean officials Jan. 29 and 31.

Seoul's action was first disclosed Jan. 29 by Myron B. Kratzer, acting assistant secretary of state for oceans, environments and scientific affairs. Acknowledging that the U.S. had discussed the matter with the South Koreans, Kratzer told the U.S. Senate's Government Operations Commmittee that the Seoul government had decided that cancellation of the purchase "was in its own best interest." He expressed hope that "it would lead other countries to realize that reprocessing facilities are not a sensible or attractive part of the nuclear power system."

The South Korean government made no official comment but a South Korean official said Jan. 31 that his country had dropped the purchase plans after the U.S. "made the strongest possible representation to the Korean and French governments. They argued that having the plant would raise the suspicion that Korea is acquiring nuclear arms. The Americans said it would raise tensions and upset the balance of power on the Korean peninsula.

The U.S. state department was said to have warned South Korea that the U.S. would withhold the required export licenses and Export-Import Bank financing for its $292 million Westinghouse atomic power reactor program if Seoul purchased the French facility. The French equipment could process used uranium from nuclear reactors and thus make South Korea less dependent on fuel from foreign suppliers.

In a related development, South Korea announced Jan. 30 that it had signed an agreement with Canada Jan. 26 for the purchase of a Candu reactor similar to that used by India in building a nuclear explosive device. The accord included a clause that said nothing provided by Canada might be used to develop and detonate nuclear devices.

Soviet-French 'hot line' A-safeguards agreement—French Foreign Minister Jean Sauvagnargues and Soviet Foreign Minister Andrei Gromyko signed an agreement in Moscow July 16 on safeguards to prevent either country from launching accidental nuclear attacks on the other. The agreement provided for the establishment of a "hot line" similar to the one set up between the Soviet Union and the U.S.

French sale of A-reprocessing plants banned—The government announced Dec. 16 that it would ban all future sales of nuclear fuel-reprocessing plants that could be used to produce plutonium for atomic bombs. This was a change from the previous French policy, which had opposed any interference with its nuclear exports.

The decision followed the French Council for Foreign Nuclear Policy's statement Oct. 11 that France would work against the proliferation of atomic weapons and was ready to study international treaties to effect the policy. Observers believed the council's statement indicated that France might ratify the non-proliferation treaty.

The ban on exporting reprocessing plants was not retroactive. An earlier French agreement to supply Pakistan with a plant remained in effect. It was reported Dec. 16 that the French government wanted to abandon the Pakistan deal but that domestic political pressure from the Gaullist Party made this action difficult. The Gaullists reportedly were extremely sensitive to the appearance of French surrender to pressure from abroad, particularly from the U.S., which had issued vigorous protest over the Pakistan agreement.

President Valery Giscard d'Estaing reportedly was hopeful that the Pakistanis would terminate the agreement, but according to a Dec. 16 report they had decided to continue with plans to purchase the nuclear plant despite U.S. objections.

An earlier deal for a reprocessing plant sale to South Korea had been canceled because of strong U.S. pressure. Agreements with South Africa for the sale of two A-plants and with Iran for the same number would remain in effect because the French felt there were enough safeguards to prevent diversion of the fuel into weapons.

The French were expected to continue to supply third-world countries with nuclear reactors, fuel and technology. Following the Dec. 16 decision, however, all irradiated fuel produced in French-supplied nuclear plants would be returned to France to be retreated so that the plutonium produced would remain in French hands.

U.S.S.R. sells India heavy water—The Soviet Union agreed to sell India 200 metric tons of heavy water for India's nuclear reactors, U.S. officials reported Dec. 8. (A metric ton is 2,024.6 pounds.)

The International Atomic Energy Agency (IAEA) was notified of the deal, and Soviet officials indicated that international inspection and safeguard procedures would be observed. The sale was considered unusual because it was a case of Soviet willingness to supply nuclear material to a country with bomb-making capability. (India had exploded its first nuclear device in 1974. Canada, formerly India's chief supplier of reactors and material, had cut off sales after the 1974 explosion and had refused further dealings with India in May after India had refused to place its nuclear facilities under IAEA inspection.)

Heavy water, used to control nuclear reactions in some reactors, is water in which the hydrogen atoms are replaced by deuterium, a hydrogen isotope twice the weight of ordinary hydrogen.

U.S. approves reactor sale to Spain.

The U.S. Nuclear Regulatory Commission June 21, 1976 approved the sale of a reactor to Spain, despite strong objections of one member of the four-man panel.

Victor Gilinsky registered the first protest of a commission member against granting a reactor-export license. He said that the deal did not provide sufficient safeguards to prevent Spain from manufacturing nuclear weapons with the reactor fuel. He had argued for a license modification to restrict Spain to using only U.S. uranium as fuel. Such a restriction would prevent Spain from reprocessing the fuel to produce plutonium, without U.S. permission. Spain had previously signed deals with U.S. companies to build reactors for generating electricity.

Ford warns on arms spread.

President Gerald Ford told Congress July 29, 1976 that a "world of many nuclear-weapons states could become extremely unstable and dangerous."

Presenting the annual report of the U.S. Arms Control and Disarmament Agency, Ford said that 20 additional countries had the technical competence and the material to build nuclear weapons. Six nations—the U.S., the Soviet Union, China, Britain, France and India—already had such knowledge and materials. Ford said that all countries supplying nuclear fuel, equipment and assistance should impose International Atomic Energy Agency standards on their sales. These standards included prohibition of all nuclear explosions, insistence upon measures to prevent sabotage and theft, and imposition of the same standards on the equipment if it were transferred to a third country.

In a related development, Fred C. Ikle, July 23 criticized U.S. nuclear aid as "too cavalier."

In an interview after he had testified before the Senate Foreign Relations Committee, Ikle said that the U.S. had sold reactors and fuel abroad with little regard for safety standards. While the situation had shown improvement in recent years, Ikle said, the U.S. had supplied reactors to countries which did not need them, such as South Vietnam and Zaire. He also charged that U.S. nuclear aid to Europe had made possible uncontrolled European sales of reactors and reprocessing equipment to less-developed countries.

Report says Taiwan secretly reprocessing nuclear fuel.

U.S. intelligence over the first six months of 1976 found that the Republic of China (Taiwan) was secretly reprocessing used uranium fuel to obtain plutonium, the Washington Post reported Aug. 29, 1976. A Taiwan Embassy spokesman in Washington Aug. 28 denied the report.

The Post said that a hot-cell reprocessing center had been used to produce about one pound of plutonium. American, Canadian and European experts were quoted as saying that this amount was far short of the minimum of 18 pounds needed for a sophisticated nuclear device, but enough to provide knowledge of plutonium handling and explosive fabrication.

The hot-cell plant had been built near Taipei with American aid and with parts obtained from countries around the world. The installation had been constructed in 1970 with the understanding that it would be subject to inspections by the Vienna-based U.N. International Atomic Energy Agency (IAEA). Taiwan was ousted by the IAEA in 1971.

Taiwan had a research reactor and four other American-built nuclear power plants in addition to the recycling plant, and hoped to acquire two additional power plants from the U.S. by 1980. But U.S. Arms Control and Disarmament Agency officials said they were stalling on the application for the two additional plants in an effort to stop Taiwan's secret uranium reprocessing, the Post report said.

Taiwan was developing its capacity to reprocess used nuclear fuel in order to produce atomic-weapons fuel, according to a U.S. Central Intelligence Agency report by U.S. officials cited Aug. 29.

Taiwan denied the report Aug. 29, saying that while it had an operational reprocessing plant, the facility could reprocess only a "tiny amount" of weapons-grade fuel for research purposes. Taiwan had not belonged to the International Atomic Energy Agency since 1972, when it was ousted in the struggle to accredit the People's Republic of China in the United Nations, but the Taipei government still adhered to IAEA safeguards and allowed inspections by the agency.

IAEA warned of NPT failure. Sigvard Eklund, director general of the United Nations International Atomic Energy Agency (IAEA), warned members at the Sept. 21, 1976 opening of its annual conference in Rio de Janeiro, Brazil, that the nuclear Non-Proliferation Treaty (NPT) would be rendered useless because a large number of nations had neither accepted the treaty nor voluntarily approved IAEA safeguards.

Eklund urged nuclear-export nations to insist that nations receiving nuclear aid adhere strictly to the safeguards in all their nuclear activities. (Imported fuel was subject to safeguards required by an exporting country, but a nation which had not signed the NPT was free to develop atomic weapons with recycled nuclear fuel. Non-signatories included Argentina, Brazil, Egypt, Israel, Pakistan, Spain and South Africa.)

At the close of the conference Sept. 27, Yugoslavia proposed an international nuclear fuel cycle pool, which would allow nations to contribute or withdraw money, technology and fuel according to their resources and needs. The pool would assist developing countries in building up their nuclear energy potential and would help developed countries by stabilizing the uranium supply.

West Germany honors Brazil agreement. Despite strong protests from the U.S., the West German government April 5, 1977 approved export licenses for blueprints of a pilot uranium-reprocessing plant and a demonstration uranium-enrichment plant to be built in Brazil.

The plants, to be constructed under a 1975 agreement between Bonn and Brasilia, would produce fissionable uranium and plutonium that could be used in nuclear weapons. The U.S. and other countries feared that Brazil would build an atomic bomb, but Brazilian and West German officials publicly denied this.

West German Chancellor Helmut Schmidt had held up the export licenses since January, when Vice President Mondale visited Bonn and told him that President Carter had serious objections to the nuclear agreement with Brazil. U.S. Secretary of State Cyrus Vance visited Bonn March 31 to apply further U.S. pressure, but Schmidt reportedly was adamant, arguing that West Germany must honor its international agreements and that Brazil would use its nuclear technology for peaceful purposes.

(Despite Bonn's public determination, a U.S. State Department spokesman noted April 8 that West Germany's initial exports to Brazil would be limited to blueprints and other data, still leaving the U.S. time to convince Bonn and Brasilia to change their plans. West German officials noted that the actual construction of the enrichment and reprocessing plants would take years, and they said Bonn was willing to confer with the U.S. on possible additional controls in future contracts with other countries.)

The U.S. pressure evoked protests in both West Germany and Brazil. In West Germany, the weekly newspaper Die Zeit charged that the U.S. wanted to destroy the agreement between Bonn and Brasilia to secure the Brazilian contracts for the U.S. nuclear industry, it was reported April 9. Count Otto Lambsdorff, economic spokesman for the Free Democratic Party, said Bonn must honor the Brazilian agreement "to uphold the international credibility of the Federal Republic of Germany as well as to insure the future of German export technology." German labor unions estimated that as many as 200,000 jobs depended on the nuclear industry.

The U.S. pressure had brought U.S.-Brazilian relations to their lowest point in more than a decade, the New York Times reported March 28. Some ranking Brazilian officials believed U.S. pressure against the nuclear deal and in favor of greater observance of human rights in Brazil were part of a larger U.S. effort to prevent Brazil from becoming a major world power, according to the Times.

Brazilian Mines and Energy Minister Shigeaki Ueki said March 26 that Brazil would carry out its nuclear energy program "at all costs." Brazil planned to rely on nuclear and hydroelectric plants because it lacked domestic oil deposits and wanted to cut its purchases of high-priced foreign petroleum.

Carter asks nuclear export curbs. President Carter April 27, 1977 proposed a bill to establish tough new controls on U.S. exports of nuclear technology and materials to prevent the spread of nuclear weapons. In a message to Congress, Carter said the bill was "intended to reassure other nations that the U.S. will be a reliable supplier of nuclear fuel and equipment for those who genuinely share our desire for nonproliferation."

Under the President's bill, nuclear materials would not be exported until the executive branch notified the Nuclear Regulatory Commission that the proposed sale "will not be inimicable to the common defense and security."

The bill also provided interim rules on exports that would be applied while foreign governments were being pressured to renegotiate existing agreements with the U.S. to incorporate the tougher new rules. The interim rules were intended to discourage the use or transfer of nuclear materials for military purposes.

Any new international agreements would have to include the new rules, according to the proposed legislation. One of the new rules required that in countries currently without nuclear weapons, "all nuclear materials and equipment" be subject to safeguards of the International Atomic Energy Agency (IAEA), regardless of whether those materials had been supplied by the U.S. No U.S. nuclear exports would be permitted to countries that did not submit to international controls. Carter said this requirement should be viewed as an interim measure, as the first

preference of the U.S. was that all nations sign the Nuclear Nonproliferation Treaty.

(If applied strictly the requirement for international controls would rule out future U.S. contracts with Israel unless Israel let experts examine its secret Dimona facility, which some observers believed was used to make nuclear exposives.

Among other rules to be included in new agreements were:

■ A requirement for IAEA safeguards on all U.S.-supplied material and equipment for an indefinite period.

■ Extension of current U.S. rights of approval on re-transfers and reprocessing to cover all special nuclear material produced through the use of U.S. equipment.

■ Exemptions from the new rules could be granted by the president "if he considers this to be in our overall nonproliferation interest."

■ U.S. nuclear supplies would be cut off to any nation that exploded a nuclear device or materially violated IAEA safeguards or any guarantee it had given under its agreement with the U.S.

In his message to Congress, Carter said that similar legislation previously introduced in Congress was too strict and did not allow the president enough flexibility. Some earlier congressional proposals, for example, would require the U.S. to stop exports immediately to foreign nations that did not promptly accept the new rules through renegotiation of existing agreements. The President's bill would allow exports to continue during negotiations about application of the rules.

The tough standards and penalties in the Carter legislation reflected the fact that the U.S. retained considerable technological leverage in the recently intensified international bargaining over nonproliferation of nuclear weapons, according to a New York Times news analysis April 28. Many nations currently using light-water nuclear power plants that were pioneered in the U.S. depended heavily on the U.S. for supplies of enriched uranium fuel for them.

The analysis noted, however, that the leading customers for U.S. nuclear equipment and material, including Britain, Japan, France and West Germany, were all investing heavily in efforts to break their dependence on the U.S. Their efforts included work on plutonium-fueled fast-breeder reactors, which used unenriched uranium to "breed" their own supplies of plutonium.

Carter plutonium ban assailed—A 41-nation conference on nuclear energy April 13 adopted a resolution opposing U.S. President Jimmy Carter's nuclear energy policy because it attempted to restrict the development of fast-breeder reactors. The resolution was approved at the end of a five-day conference in Persepolis, Iran attended by 500 scientists, government officials and representatives of the nuclear industry.

The conference had been called by the American Nuclear Society and similar organizations in Japan and Europe. Among the nations represented were the U.S., the Soviet Union, France, West Germany and Japan.

The resolution said that most countries looked to nuclear power as the only means to energy independence. For those countries without large uranium resources, this independence could come only through the use of breeder reactors, it said.

The resolution also contended that the Carter restrictions unilaterally abrogated the section of the Nuclear Nonproliferation Treaty that promoted the free flow of nuclear knowledge. This, the resolution said, weakened the confidence of other nations in U.S. promises to provide adequate uranium fuel supplies as alternatives to plutonium. [See above]

In other reaction, officials of the International Atomic Energy Agency (IAEA) in Vienna believed the Carter policy could not be successful, according to the Los Angeles Times April 17. They said that not a single exporter of nuclear technology had indicated support for it.

Sigvard Eklund, IAEA director general, said the policy could "disrupt the atmosphere of mutual trust and confidence that has been built up during the last 25 years." Another IAEA official said it "is a fallacy to suppose that because plutonium and reprocessing are unnecessary and uneconomical for the U.S., they are also unnecessary and uneconomical for the rest of the world."

On the U.S. domestic front, Robert E. Kirby, chairman of Westinghouse Electric Corp., told the company's annual meeting April 27 that Carter's proposed curb on fast-breeder reactor research was "most unwise." He said breeder reactors promised energy resources equivalent to at least three times those represented by Middle East oil reserves.

Westinghouse, a diversified manufacturer of industrial and electrical products, had been designing the nation's first large-scale breeder research plant. Its construction on the Clinch River at Oak Ridge, Tenn. had been postponed by Carter.

Kirby said "the connection between the breeder program and possible nuclear weapons proliferation is so thin as to be insignificant." He warned that halting breeder research would "remove the U.S. from any voice in the development of international controls of this important new technology."

U.S. Sen. Frank Church (D, Ida.) echoed that theme May 2 when he called the President's policy "a formula for nuclear isolationism" that would "reduce, not enhance, U.S. influence in shaping worldwide nuclear policy."

Church advocated effective IAEA supervision of all uranium enrichment and reprocessing plants, saying the "agency should own all the nuclear fuel in its global system, be responsible for the fuel's security and accountable for its use."

Nuclear conference held in Salzburg. Most of the more than 60 nations represented at the International Conference on Nuclear Power and Its Fuel Cycle in Salzburg, Austria May 2-13, 1977 showed no willingness to retreat from their plans to use nuclear power or use plutonium fuel to supply that power. U.S. President Jimmy Carter previously had appealed for deferral of the use of plutonium worldwide to help curb the spread of nuclear weapons.

Attending the conference, which was sponsored by the International Atomic Energy Agency (IAEA), were some 2,000 government delegates, industrial representatives and academic specialists. Among the issues discussed were the difficulties of reactor safety, the spread of nuclear weapons, the development of new technologies and the risks of wide use of nuclear power.

The issue of nuclear nonproliferation, another primary concern of President Carter, was a major one at the conference. Much attention was paid to making nuclear bomb-grade fuels inaccessible, either by limiting fuel-processing technology or by imposing international controls. A problem debated by uranium enrichment specialists, for example, was that new methods of enrichment might make production of bomb-grade uranium too simple. The making of bomb-grade materials currently was dependent on relatively slow, highly sophisticated processes.

(Raw uranium consisted mostly of the isotope Uranium-238, which could be used neither as a fuel nor in a bomb, and about .7% of the isotope Uranium-235. Uranium enrichment was the process by which the concentration of Uranium-235 was increased. Enrichment to a 3%–5% concentration of Uranium-235 was sufficient for reactor fuels. Bomb-grade uranium had approximately a 90% concentration of Uranium-235.

Two methods of enrichment currently in use were gaseous diffusion and centrifugation. Both methods relied on the slight difference in weight between Uranium-235 and Uranium-238 in order to separate the two isotopes.

Uranium-238, when bombarded with neutrons, was converted into radioactive Plutonium-239, which could be used as a reactor fuel or a nuclear explosive. The chief motive for using plutonium as a fuel was that it could be derived in a fast-breeder reactor from the otherwise useless Uranium-238. Since most natural uranium was in the form of Uranium-238, use of plutonium-fueled breeders would greatly extend the life and usefulness of existing uranium reserves. A breeder reactor could create more plutonium from Uranium-238 than it used as a fuel.)

French delegates to the conference announced May 6 that France had developed a new method of uranium enrichment that could not practically be used to make bomb-grade material. The process involved the separation of Uranium-235 and Uranium-238 by means of chemical exchange. Two safeguards against proliferation cited by the French were the danger of an explosion in the plant if an attempt were made to reach 90% enrichment and the fact that, in any case, it would take about 30 years to achieve that level of enrichment. Other scientists, familiar with the theory but not the details of the French technique, said such a process would involve a very large plant and a very large uranium inventory. Both of these factors, they said, would make it difficult for the process to be commercially competitive.

Another major question for the future of nuclear energy was that of public acceptance and assessment of the risk. Throughout the 12-day meeting, leaders of national atomic energy programs conceded that public hostility was their chief handicap.

A central issue in public acceptance was the difficulty of accurately assessing the risk of nuclear energy use. At the conference, delegates cited a 1975 report by the U.S. Nuclear Regulatory Commission that estimated the chances of a major accident causing several thousand deaths at one in a billion for each year of operation of a power plant. Dr. Leonard D. Hamilton of Brookhaven National Laboratory and Dr. A.S. Manne of Stanford University presented a paper in which they said coal was more hazardous than uranium as an energy source. Using estimates of mortality attributable to air pollution and including hazards of mining and transport of fuel, they blamed coal for between 1,900 and 15,000 deaths in 1975, compared with between 18 and 42 deaths due to nuclear energy programs.

In other conference developments:

■ In his nation's first public response to President Carter's curtailment of the U.S. breeder-reactor research program, a delegate from the Soviet Union told the conference May 9 that "plutonium must be used." The Soviet Union would continue its development of breeder reactors, he said.

■ An article by IAEA staff member Morris Rosen, distributed to the delegates May 3, said that developing nations might be buying and installing nuclear power plants that were less safe than those operating in supplier nations. Rosen's article, which was widely discussed, pointed to the danger of earthquakes in developing countries with nuclear programs; to "sub-minimal" local watchdog organizations, and to the failure of nuclear suppliers to update design and safety features of exported reactors.

■ Dr. Hannes Alfven, winner of the Nobel Prize and a member of Sweden's Atomic Energy Commission, told the conference May 9 that nuclear energy should be abandoned. He gave four main objections: that it was associated with the production of radioactive substances; that the link between nuclear energy and weapons was so strong that "we have to accept both or neither"; that it was becoming difficult to support the claim that nuclear energy was cheap, and that there were "several other and more attractive" ways of solving the energy problem. Earlier in the day, a leading U.S. proponent of nuclear energy use, Dr. Hans Bethe, also a Nobel Prize-winner, had said nuclear energy was "a necessity, not an option."

'Counterconference' held—A group of about 100 scientists, environmental specialists and antinuclear activists from 20 countries May 1 condemned the use of plutonium as a source of energy and called for a stronger fight against conventional nuclear power plants. The group, which called itself the Salzburg Conference for a Non-Nuclear Future, met in Salzburg from April 29 to May 1. It advocated international efforts to develop such alternative technologies as the large-scale harnessing of solar energy.

The conferees reportedly worked out a code to regulate the export of nuclear equipment for peaceful purposes. But meetings on additional safeguards were inconclusive. Sources said this occurred in part because it was felt that U.S. policy was not sufficiently formulated to make definitive reaction possible.

Carter approves uranium shipments. President Carter had approved the shipment of approximately 827 pounds of highly enriched uranium to five countries for use in nuclear research reactors, it was reported May 6, 1977.

Applications for shipments of about 2,-866 pounds of enriched uranium had been pending for as long as 18 months. Officials in Western Europe had been complaining for nearly a year about the delay, asserting recently that without swift U.S. action, some of their nuclear facilities would have to cut back operations by summer. Some U.S. allies also had complained that Carter's slowness in approving the applications seemed to contradict his assertion that the U.S. would remain a reliable uranium supplier.

The shipments approved by Carter were destined for Belgium, Canada, Japan, the Netherlands and West Germany. In addition, the State Department had cleared the shipment of as much as 496 more pounds of high-grade uranium in amounts too small to require presidential review. All the applications still had to be reviewed by the Nuclear Regulatory Commission.

Canada, U.S. agree on uranium sales— Canada and the U.S. Nov. 15, 1977 signed an agreement governing sales of Canadian uranium by the U.S. The move opened the way for the U.S. to export enriched Canadian uranium to Europe and Japan.

Under the terms of the agreement, the U.S. would consult with Canada before selling abroad any Canadian uranium that had been enriched in the U.S. Countries that wanted to reprocess the uranium to obtain plutonium would be required to ask permission only from the U.S. and not from both the U.S. and Canada.

(Enriched uranium was uranium that had been refined to contain a greater proportion of fissionable uranium than in uranium ore. After being used in a nuclear reaction, it could be reprocessed to obtain plutonium, a fuel used in nuclear weapons.)

Nuclear exporters OK safeguard pact. Delegates at the Conference of Atomic Energy Suppliers in London initialed a pact Sept. 21, 1977 to institute safeguards on sales of nuclear energy and technology to other countries. The pact was aimed at containing the spread of nuclear weapons technology.

Efforts to halt the spread of nuclear weapons had been hampered because not all nuclear-exporting nations required their customers to observe international safeguards against making weapons. Under the terms of the London agreement, nuclear exporters would require customers to: pledge not to develop or detonate atomic weapons even for peaceful purposes; permit inspection of nuclear facilities by the International Atomic Energy Agency (IAEA); show that nuclear installations were protected against sabotage and theft; agree to IAEA controls on resales of nuclear material to third countries, and pledge not to duplicate nuclear facilities purchased under the London controls.

The agreement had been reached in principle in 1976 and had been reaffirmed earlier in 1977. The final pact clarified the language of the agreement and added provisions for dealing with violations. The text would be submitted for IAEA approval. The countries involved in the London conference were: Belgium, Canada, Czechoslovakia, East Germany, France, Great Britain, Italy, Japan, the Netherlands, Poland, Sweden, Switzerland, the U.S.S.R., the U.S. and West Germany.

The pact had been confirmed in the U.S. Feb. 23, 1976 when Fred C. Ikle told the Senate Foreign Relations Subcommittee on Arms Control that the agreement covered not only guidelines for preventing nuclear exports from being turned into atomic weapons, but also follow-up efforts to improve safeguards. A State Department official said later that day that Ikle was prevented from giving more complete testimony because several countries, notably France, had placed restrictions in the understanding on the disclosure of details, including the identity of the participants.

Export of nuclear data halted. Chancellor Helmut Schmidt announced June 17, 1977 that West Germany would halt the export of nuclear technology that could be used to construct atomic weapons. He qualified the policy by reaffirming that existing contracts for atomic projects, such as the controversial sale of reprocessing equipment to Brazil, would be continued.

Schmidt's statement came at the end of a two-day visit to Bonn by French President Valery Giscard d'Estaing to discuss energy policies. The West German policy was similar to the French ban on sensitive nuclear exports announced in December 1976. The French also had said that existing contracts, such as their planned sale of reprocessing equipment to Pakistan, would be unaffected by the export ban.

The new West German policy was part of the stepped-up effort by industrialized nations to stop the spread of nuclear weaponry. Although the ban did not apply to the Brazil contract, it meant that similar contracts with Iran or other nations would be impossible. The policy also banned such sales even if other nuclear exporters continued to export the technology.

The West German contract with Brazil called for the construction of eight nuclear plants at a cost of $5 billion. Construction of the plants was not controversial, but the contract also called for a reprocessing plant that could be used to produce plutonium for atomic weapons.

U.S. President Carter had clashed with Schmidt several times in the past, on the export of advanced nuclear facilities to Brazil.

Carter & Japanese disagree. President Carter met with Japanese Premier Takeo Fukuda in Washington March 21–22, 1977. One of the subjects discussed was atomic energy.

The two leaders failed to agree on Japanese plans to build a nuclear processing plant that would use U.S.-supplied nuclear fuel, a principal topic of discussion March 22. The plant was scheduled to start test operations in the summer. Carter told Fukuda that the facility was unnecessary because it would be economically wasteful and would contribute to the spread of technology that produced nuclear weapons, Japanese officials said. Fukuda was said to have argued that the plant would provide Japan with a source of energy that was an alternative to oil. The premier also noted that Japan had signed the Nuclear Nonproliferation Treaty in 1976 with the understanding that it would not be hampered in its development of nuclear energy for peaceful purposes.

U.S. OKs Japan spent-fuel export—The U.S. had agreed to a resumption of Japan's transfer of used nuclear fuel to Great Britain for reprocessing, the Japan Atomic Power Co. reported March 28. A shipment scheduled for March 29 would be the first since October 1976, when former President Ford asked the Japanese government to suspend such shipments. Ford had called for a three-year ban on exports of spent nuclear fuel and uranium enrichment technology.

U.S. plutonium policy scored—Japanese Foreign Minister Iichiro Hatoyama said March 30 that the U.S. proposal to ban nuclear waste reprocessing violated the Nuclear Nonproliferation Treaty, which guaranteed the peaceful uses of nuclear energy. Premier Takeo Fukuda had discussed Japanese plans to build an atomic processing plant with U.S.-supplied nuclear fuel at a meeting with President Carter March 22. Carter April 7 formally proposed a ban on the use of plutonium in commercial nuclear power plants in the U.S.

Speaking to a parliamentary budget committee, Hatoyama warned that the U.S. might block Japan's program for peaceful use of nuclear energy. At the same time, Japanese government sources disclosed March 30 that the U.S. had formally apprised Japan of its basic position on the matter.

Japanese concern over American nuclear energy policy was reported again April 2. An official of the government's nuclear industry division said it was "difficult to understand the American philosophy." Anticipating the Carter program, the official was further quoted as saying, "We think America will defer reprocessing indefinitely and ask other countries to follow the same policies . . . even if it's impossible."

A Japanese diplomat complained that his country had "followed U.S. guidelines on nuclear policy" for 20 years. "Now," he said, "you are saying you made a complete mistake . . . but it's too late."

Tokyo officials feared that President Carter's proposed curbs would further undermine Japanese public confidence in a program that was already controversial. Advocates of nuclear power had to fight for each project site because of environmentalists' opposition. As a result of this resistance, targeted power production projects were falling 50% behind schedule.

Japan starts up first breeder—Japan's first experimental fast-breeder reactor April 24 reached criticality, or the point at which a nuclear reaction was virtually self-sustaining. It was the first time Japan had ever used plutonium to spark a critical nuclear reaction. The exercise made Japan the fifth nation to breed plutonium for peaceful purposes. The other countries were the U.S., the Soviet Union, Great Britain and France.

The reactor, named "Joyo," was built by the government's Power Reactor and Nuclear Fuel Development Corp. at a cost of $100 million. It was the first of two planned prototypes. There were 12 similar reactors in the other four countries.

Only days before, on April 21, the Tokyo government had announced it would seek a "common front" with Great Britain, France and West Germany in an effort to have proposed U.S. curbs on the use of plutonium modified.

U.S., Japan agree on nuclear reprocessing—The U.S. and Japan Sept. 12 signed an agreement permitting Japan to open a nuclear reprocessing plant.

The agreement between the two countries allowed Japan to open its newly built Tokai Mura reprocessing plant for an experimental two-year period. During that time, Japan agreed to conduct intensive research to determine if alternative reprocessing methods could be developed—methods that would produce useable reactor fuel from spent fuel, but would not produce weapons-grade material.

U.S. approval of the Japanese plant was required because the U.S. supplied Japan with the uranium it used as the basic fuel for its power plants, and the U.S. retained jurisdiction over the uranium even after it had been used to fuel the plants.

Under the agreement, Japan would not fully reprocess the spent fuel. Plutonium nitrate would be produced, but it would not be converted into plutonium oxide, which was the next step along the reprocessing route.

Nuclear Export Safeguards Made Final.
The U.S. State Department said Jan. 11, 1978 that 15 nations—including the U.S. and the U.S.S.R.—had agreed on a set of rules for the export of nuclear technology that were designed to prevent the spread of atomic weapons. The 16-provision code, which had been negotiated at the Conference of Atomic Energy Suppliers in London in 1977, was being submitted to the International Atomic Energy Agency (IAEA) in Vienna, the State Department said.

The code, in its more important provisions:

■ Required nuclear export deals to include "governmental assurances" that the exports would not be put to the purpose of weapons production.

■ Stated that in the case of a dispute between a supplier country and a customer over safeguards, other suppliers would not undercut the original supplier country's negotiating position by offering their own exports.

■ Required supplier and customer countries to make security arrangements to protect nuclear materials and technology against theft.

■ Made export sales conditional on the customer country's agreement to accept international safeguards, including periodic inspections of facilities included in an export agreement.

■ Stipulated that customer countries could not reexport nuclear technology or materials except in accordance with the code of safeguards set by the supplier countries.

Another provision required supplier countries to show restraint in selling "sensitive facilities"—that is, technology that could be turned to weapons production with relative ease. State Department officials said the provision would ban the sale of reprocessing equipment, which could be used to extract plutonium from the spent fuel of conventional atomic reactors. The ban, the officials said, would take effect in the future, and would not

affect two reprocessing deals that were already underway: an agreement between West Germany and Brazil, and a deal still being negotiated between France and Pakistan.

(It was not known whether the State Department interpretation of the provision was shared by France and Germany.)

The code announced Jan. 11 did not require so-called "full-scope" safeguards, under which a country importing nuclear technology would have to agree to inspection of all of its nuclear facilities, no matter how or when acquired. However, Joseph Nye, a State Department official concerned with nuclear proliferation, Jan. 11 called the public agreement on the code "a move forward toward full-scope safeguards."

In addition to the U.S. and the Soviet Union, the other 13 nuclear supplier nations who were parties to the safeguards agreement were: France, Great Britain, Japan, West Germany, Canada, Poland, East Germany, Czechoslovakia, Belgium, the Netherlands, Italy, Sweden and Switzerland.

Nuclear Export Controls Signed.
President Carter signed legislation March 10, 1978 imposing stricter controls on the export of nuclear technology and fuel. The main purpose of the bill was to prevent the spread of nuclear weapons.

The bill provided that the U.S. would end nuclear exports to any country without nuclear weapons that developed such weapons. Exports would also be cut off, under the bill, if a country violated international atomic safeguards.

The legislation barred countries from re-transferring without U.S. consent fuel they had received from the U.S. Consent would also be required for reprocessing spent nuclear fuel.

The bill included provisions designed to assure other countries that the U.S. would be a reliable supplier of nuclear fuel and equipment, provided the countries abided by the safeguards against weapons proliferation. The U.S. hoped that such assurances would persuade other countries to refrain from using reprocessing technology or building breeder reactors.

(Breeder reactors and reprocessing

technology were particularly attractive to countries without domestic supplies of uranium. They would free the countries from reliance upon a supplier country. However, those technologies—in the form in which they had been most developed—would also produce plutonium, which could be used to make atomic weapons.)

The U.S. nuclear industry had opposed the export-controls legislation. The industry contended that foreign countries would simply turn to other nations for their nuclear supplies, thus evading the controls and diminishing U.S. exports and profits.

Carter, when he signed the bill, called it a "major step forward in clarifying our own nation's policy."

Carter said he felt "very strongly that we should continue to use in an increasing way atomic power. . . ." He added, however, that light water nuclear reactors were adequate for the present and that it would be a mistake to build a production model breeder reactor.

As a consequence of the bill's export controls, Carter said, "Some of our friends will have to readjust their policy." Observers interpreted the remark as a reference to France and other nations that had not signed an international treaty to prevent the spread of nuclear weapons.

The bill had cleared Congress Feb. 9 when the House of Representatives approved it by voice vote. The Senate had passed it Feb. 7 on an 88-3 vote.

U.S. Nuclear Export Policy Disputed—A major dispute between the U.S. and the European Community over the renegotiation of uranium export contracts was discussed April 7-8 at the Copenhagen meeting of the EC heads of state. The dispute was one of a number of EC-U.S. conflicts that had burdened relations between the Western allies. [See above]

The conflict centered on the new Nuclear Nonproliferation Act passed by the U.S. Congress in March. The act obliged the U.S. to renegotiate contracts under which the U.S. supplied enriched uranium to foreign nations. The new measure was part of the nuclear safeguards policy designed to insure that the U.S. could veto the re-exportation of the material or its use for reprocessing into weapons-grade plutonium.

The close connection in time between the uranium export dispute and the controversy over U.S. handling of the neutron bomb issue reportedly had convinced European leaders that U.S. policy was unpredictable and that the Carter Administration was particularly weak.

The French government April 4 had blocked the EC from replying to the U.S. demand for renegotiations of the uranium export contracts. The French argued that the U.S. had agreed in 1977 to permit existing contracts to stay in effect for two years. This made the new U.S. demands unjustified, according to the French.

The other EC nations backed the French position by allowing the April 9 deadline imposed by the U.S. on the issue to lapse. U.S. shipments of nuclear material were supposed to stop by that date to any nation that had not agreed to submit to the new laws.

British Foreign Secretary David Owen said April 8 that talks on the subject would be held with the U.S. at a later date. He said they would not be formal negotiations, because the EC nations did not recognize the U.S. right to impose new contract terms after an agreement had been signed.

Western Europe relied upon the U.S. for over half of its supply of nuclear reactor fuel and most of the enriched uranium for nuclear research. EC nations were concerned that the new U.S. uranium export policy would give the U.S. power over all their nuclear programs, particularly exports and the construction of fast-breeder reactors.

A measure of EC displeasure with U.S. policies was reflected in unconfirmed reports that West Germany would seek supplies of enriched uranium from the Soviet Union. West German Chancellor Helmut Schmidt and French President Valery Giscard d'Estaing reportedly had discussed the issue during their meeting April 2 at Rambouillet, near Paris. The subject of the uranium purchases reportedly would be raised at an upcoming meeting in Bonn between Schmidt and Soviet President Leonid Brezhnev.

U.S. OKs last-minute uranium exports—
The U.S. Nuclear Regulatory Commission allowed the export of 1,000 pounds (455 kilograms) of enriched uranium to European Community nations, it was reported April 17. The shipment was approved just before an April 9 deadline, when stricter U.S. export rules went into effect.

According to press reports April 17, the action was taken to ease tensions between the EC and the U.S. over the U.S. intention to renegotiate all export contracts for enriched uranium. Further shipments of enriched uranium to Europe were in doubt because Euratom, the EC's agency for nuclear affairs, had refused to agree to the U.S. demand to renegotiate the contracts.

The 1,000 pounds of uranium were destined for nuclear facilities in West Germany, France and Denmark.

U.S.-Soviet A-safety pact set. The U.S. and the U.S.S.R. had signed a protocol on cooperation in the field of nuclear power plant safety, as reported Feb. 22, 1978 in the Journal of Commerce. The report said the protocol would serve as the basis for a future agreement on U.S.-Soviet exchanges of information and delegations to study atomic energy safety in both countries.

The Journal reported that a U.S. delegation on nuclear power safety had visited the U.S.S.R. Feb. 5–18 and had been favorably impressed with Soviet work in safeguarding nuclear power plants from accidents. The six-member delegation said it had visited several atomic power stations and had discussed design and safety procedures with Soviet experts.

Country-by-Country Survey

AUSTRALIA

U.S.-Australian safeguard views. Prime Minister Malcolm Fraser told the Australian Parliament March 23, 1977 that he had exchanged letters with President Carter on international safeguards for nuclear facilities. Fraser also said Australia would probably become a major supplier of uranium in the world market if the government authorized uranium development, but that such development would be subject to strict controls.

Fraser's letter to Carter, written Feb. 11, pledged Australia to stringent control over uranium exports. It was intended to align Australia's policies with those of the U.S., Canada and other uranium producers.

Carter's reply to Fraser included a proposal to guarantee fuel supplies to nations that accepted non-proliferation safeguards and did not acquire sensitive nuclear facilities that could be used to produce weapons. Carter said that Australia's potential as a major supplier of uranium gave the country a particular interest in collaborating with like-minded countries on supply policies.

The Australian government was under continuous pressure from other nations that wanted its uranium deposits developed and exported. A delegation of U.S. utility executives held talks Feb. 24 with the government and the mining industry on the subject. The executives said the U.S. nuclear power industry would go ahead "with or without Australian uranium." However, the group was clearly anxious to encourage the pro-uranium faction in Australia by holding up the prospect of Australia being left out of the lucrative uranium market.

The Philippines had sent a representative to inquire about the purchase of uranium, it was reported March 8, and a European Community (EC) delegation held talks March 14 and 16 on Australian mineral resources. In particular, the EC representatives were interested in uranium, coal and natural gas.

In a related development, the Australian uranium company, Queensland Mines Ltd., told two Japanese power utilities March 24 that it would be unable to honor export contracts in 1977 for 400 tons of uranium. The company said that the government had prevented development of the rich Narbarlek deposit in the Northern Territory and had refused to supply the uranium from the government stockpile. The government had promised foreign utilities that contracts signed in 1972 would be honored whatever the final decision on full-scale development of the reserves.

There had been a long debate in Australia over the development of the nation's uranium reserves. Trade unions, environmental groups and others had argued that Australia could help cause a nuclear war by adding to the world's uranium supply. An embargo had been imposed on uranium exports in 1972, which was partly lifted to honor contracts signed before 1972.

Fraser rejects uranium referendum— Prime Minister Malcolm Fraser rejected a union proposal Sept. 7 to hold a national referendum on uranium mining and export. Fraser said the unions wanted veto power over government policy and that he would not give in to them.

The idea of a referendum on the uranium ban had been trade union and Labor Party policy since December 1976. Robert Hawke, president of the Australian Council of Trade Unions (ACTU) had brought it up again Sept. 6 after the government had lifted the ban.

Hawke said that the trade unions would cooperate fully in the development of uranium if the referendum approved it. However, if the government refused to hold a national vote on the issue, the unions reportedly would consider putting a ban on all mining, processing and transport of uranium.

The annual ACTU congress voted 493–371, Sept. 15 in favor of a national referendum on Australia's uranium policy. The ACTU threatened a work boycott on any uranium-connected project if the government failed to agree to hold a vote on the issue within a year.

Hawke reportedly worked hard to prevent the passage of a more radical proposition that backed the adoption of the Labor Party's policy of an indefinite moratorium on uranium development. Hawke argued against the position, saying that the trade union movement had no more right than the government to impose a policy on the Australian people. He

warned, however, that if the government refused to hold a national vote on the uranium issue, there would be "troops on the wharves and bloodiness."

Uren leads anti-uranium drive—Tom Uren, deputy leader of the opposition Labor Party, had become the leading spokesman for the anti-uranium movement in Australia, it was reported Oct. 18. Commentators compared the anti-uranium campaign to the protest movement during the Vietnam War. The uranium issue also had been raised repeatedly by both the government and the opposition as a possible election issue.

Uranium mining bills. The Australian Parliament passed six bills May 31, 1978 outlining the requirements for uranium mining. The government's action permitted mining companies to begin preparations for the development and export of Australia's extensive uranium deposits.

The new bill provided for environmental protection of the uranium mine sites, land rights for the aborigine owners of the sites and nuclear safeguards for mining and export of radioactive materials.

Uranium policy protested—Demonstrations took place in Australia's major cities April 1 in protest against the government's decision to permit the nation's uranium reserves to be mined and exported. The demonstrations in Sydney, Brisbane, Adelaide, Perth and Melbourne were peaceful.

Scientists urge ban on uranium mining & export—In a statement printed Jan. 3, 1977, a group of 200 Australian scientists had urged the government to ban the mining and export of uranium. They asked that the government instead formulate a comprehensive energy conservation policy and an alternative energy development program.

Charles Birch, professor of biology at the University of Sydney and one of the group's leaders, said, "The mining and export of Australian uranium will substantially increase the risk of nuclear war and the risk of a major catastrophe in nuclear power plants."

The statement itself, which was published in the Australian weekly, National Times, said that the dangers of nuclear terrorism and the problems of radioactive waste disposal outweighed the benefits of nuclear power.

Robert Robotham, radiation protection officer at Melbourne University and another of the signers, said those who signed the statement, including biologists, engineers, chemists, physicists and social scientists, had different reasons for their opposition to nuclear development. "For instance," Robotham said, "the geologists would be concerned about deposits of radioactive wastes in the earth's crust."

Australia was believed to possess almost 380,000 tons of uranium, which was 20% of the world's known reserves. (The extent of the uranium reserves in the Soviet Union and China was unknown in the West.) Economic factors and opposition from conservationists, unions and church groups had delayed development.

Sir Charles Court, premier of Western Australia was reported Jan. 11 to have criticized the 200 scientists.

Court argued that the scientists' "negative" attitude toward uranium mining would cause Australia to be left behind by the rest of the world in the "inevitable development" of nuclear power sources. He also warned that young Australian scientists would have to leave the country to pursue their careers if they chose fields involving nuclear power.

Fraser names Fox nuclear envoy. Justice Russell Fox was appointed Australia's ambassador-at-large for nuclear affairs Oct. 6, 1977 by Prime Minister Malcolm Fraser.

In his new position, Fox would advise the government on safeguards agreements with nations receiving Australian uranium.

Uranium mining ban lifted. Prime Minister Malcolm Fraser had told the House of Representatives Aug. 25, 1977 that the government had decided to lift its four-year-old ban on uranium mining and export. The move was criticized immediately by the opposition Labor Party,

trade union leaders and other critics of nuclear power development.

Fraser emphasized that the safeguards placed on the uranium mines and exports would be "stricter and more vigorous than any adopted by any country to date." He also promised that a national park would be created in the Northern Territory where most of the deposits were concentrated.

The government policy closely followed the recommendations of the Fox Commission. The Fox Commission's report had suggested that the government approve uranium exports, but that development be tightly controlled and that strict control over the use of the exports be enforced.

Uranium export guidelines set—Before the Fox report was issued, Prime Minister Malcolm Fraser May 24 announced restrictions would be imposed on any exports of Australian uranium. The controls were similar to those already implemented by the U.S. and Canada.

The restrictions provided that:

—Sales would be made only to nations that had signed the Nuclear Nonproliferation Treaty and accepted International Atomic Energy Agency safeguards.

—Buyers would have to sign separate bilateral agreements with Australia governing the use and control of Australian-supplied uranium. Nuclear material could not be diverted to military or explosive use.

—Australia would have the right to stop the transfer of nuclear material to a third party.

Fox Commission report—The government commissioned report, headed by Justice Russell Fox, had been released May 25. The report said that development of the Ranger uranium mine in the Northwest Territory should be permitted only if the "best environmental protection technology available anywhere in the world" were employed.

The report was the second and final one by the three-member Ranger Uranium Environmental Inquiry Commission. In presenting it, Fox said the report should not be regarded as advocating or opposing the mining of uranium in the Arnhem Land aboriginal reserve. However, the report said that if the government were to permit uranium mining there, it should do so only in accordance with the recommendations of the commission.

Among those recommendations were that:

■ The deposits should be mined one at a time, beginning with the Ranger field and followed by Pancontinental Mining Ltd.'s Jabiluka field.

■ Development of the Jabiluka deposit, reputedly the world's largest, should be subject to a separate new environmental impact study.

■ The Koongarra deposit of Noranda Australia Pty., a subsidiary of Toronto-based Noranda Mines Ltd., should not be developed "for the present" because of the danger of serious pollution of potential national park land.

No recommendation was made concerning Queensland Mines' Nabarlek deposit, but the panel concluded that development of a mine there would have little effect on the region's ecology.

The commission also recommended that several steps be taken to protect aborigines in Arnhem Land from the effects of development:

■ A national park should be established surrounding but excluding the Ranger and Jabiluka mines. Land rights to a large part of the park should be turned over to the aborigines, who also should receive royalties from any uranium mines.

■ Aborigines should participate in the planning and management of the park.

■ Strict limitations should be imposed on the size of any mining towns and on the use of Arnhem Land by tourists. (This recommendation arose from the commission's opinion that the "greatest threat to the environment and particularly to the welfare, well-being and culture of the aboriginal people may prove to be a large white population which the mining ventures will bring.")

Nuclear foes protest—Demonstrators in Sydney threw dirt, bottles and other objects at Fraser Aug. 26 during a protest against the government's uranium decision. Other demonstrations against the policy were planned in Australia's major cities.

Labor Party leader Gough Whitlam denounced the lifting of the uranium ban, it was reported Aug. 26. He said the government had "jumped on the gravy train of a technology that will have a maximum life of 50 years and will produce toxic wastes that will endure for a quarter of a million years."

Tom Uren, deputy leader of the Labor Party, led an anti-uranium demonstration outside Parliament when Fraser delivered his address lifting the ban. Uren warned mining companies to be prepared to lose their investments in uranium mines if they proceeded to open mines as long as the Labor Party rejected uranium development. He said the next Labor Party government would repudiate contracts signed by Fraser's government.

Bob Hawke, head of the Australian Council of Trade Unions, asked Aug. 26 for a referendum on the decision to lift the ban. Hawke said a full public debate was needed on the issue because it had split the trade union movement and had divided the Australian people.

Fraser told a news conference after his speech that negotiation of export treaties would begin as soon as possible. He said the contracts would be written to give the investors as much security as possible against repudiation by a future Labor government.

The government's program for uranium development was based on the agreement made by the former Labor government under Gough Whitlam with Peko-EZ, an Australian mining company. Peko-EZ was a joint holding of two other Australian mining companies, Peko Wallsend Ltd. and EZ Industries Ltd.

Aborigines OK uranium mining. The government and representatives of aboriginal communities in the Northern Territory signed an agreement Nov. 3, 1978 opening the way for work to begin at the Ranger uranium-mining site.

The accord provided for the aborigines to receive royalties estimated at about U.S.$11 million per year.

Government officials said they expected work to begin at the site as soon as the wet season ended in April.

The Ranger site was thought to hold 110,600 tons of uranium, with a value of around US $4.5 billion. About 2,500 to 3,000 tons of uranium were expected to be produced annually.

The royalty the agreement provided for the aborigines would be 4.25% of the value of the uranium oxide produced.

Finland signs uranium accord—Finland and Australia signed a pact July 20, 1978 setting safeguards for uranium exports to Finland. The accord permitted Finland to purchase uranium from Australia for use in its nuclear power facilities.

The agreement required Finland to seek Australian consent to enrich uranium supplied by Australia or to reprocess it for reuse as fuel. It also required Australian consent for Finland to re-export any uranium to a third country. The re-export clause was to insure that the uranium would be sent to a country that had agreed to international controls.

The accord was signed in Sydney by Deputy Prime Minister John D. Anthony and Finnish Acting Trade Minister Paul Paavela. It contained a pledge that uranium supplied by Australia would not be used for military purposes. It included a condition imposed by Australia that affirmed Australia's right to halt shipments if Finland violated any of the terms of their bilateral agreement. Australia also reserved the right to cancel shipments if safeguards of the International Atomic Energy Agency were violated by Finland. The international agency would monitor Finland's compliance with the safeguards policy.

Foreign Minister Andrew Peacock called the signing of the Finnish agreement "an important first step in the establishment of a network of bilateral agreements between Australia and other countries." According to press reports July 21, the Australian government believed that the agreement to supply Finland with uranium was in line with U.S. policy, which restricted reprocessing of the material.

The Finnish nuclear program had been developed in cooperation with the Soviet Union. It consisted of one plant in operation, one that was about to be activated and two under construction.

AUSTRIA

Voters reject A-plant. Voters in Austria Nov. 5, 1978 narrowly rejected a plan to open the country's first nuclear power plant. The opposition vote was just over half: 50.5%. The turnout for the referendum—the first in Austria since World War II—was about 64% of the country's five million eligible voters.

The plant, located at Zwentendorf, on the Danube northwest of Vienna, had been virtually finished more than a year earlier, but its commissioning had been delayed. About $560 million had been spent on the facility.

Opponents questioned how the radioactive waste from the plant would be disposed. They also claimed the plant was unsafe because it had been built near an earthquake fault. Government experts said the danger from an earthquake was negligible.

Although the referendum dealt specifically with the Zwentendorf plant, the vote emerged as a political contest between Chancellor Bruno Kreisky's Socialist Party government and the opposition parties. Kreisky's government was in favor of nuclear power and the main opposition, the People's Party, also favored nuclear power in principle. But the leadership of the People's Party, citing safety considerations, called for a vote against the plant in the referendum, even though it had been in office when plans for the nuclear facility were approved.

Kreisky told journalists Oct. 23 that he regarded the referendum as an indicator of confidence in his government and might resign if the vote went against him.

When the referendum produced a majority against the plant, Kreisky called it a "shocking defeat" for his government.

However, Kreisky decided not to resign. Hertha Firnberg, the minister of scientific research, said Nov. 6 that the Socialist Party leadership had expressed its "unconditional trust" in Kreisky.

Although the referendum was not constitutionally binding, Firnberg said the government would respect the vote and not put the plant into operation.

Firnberg added that the Socialist Party was planning a major campaign to make nuclear power more acceptable to the voters. About 64% of the energy currently consumed in Austria came from foreign sources, and the dependence on foreign sources was expected to grow. Analysts had argued that nuclear power was vital for the future health of the Austrian economy. All the countries surrounding Austria had nuclear power programs under way.

FRANCE

Bombs hit nuclear station. Two bombs exploded at a nuclear power station being built at Fessenheim, about 45 miles south of Strasbourg, May 3, 1975. An anonymous caller telephoned newspapers to warn of the impending explosions, saying it was the work of the "Meinhof-Puig Antich group," apparently named after the German anarchist Ulrike Meinhof and a Spanish nationalist leader executed in 1974, Salvador Puig Antich. French officials said that the blasts occurred in a building near the nuclear reactor, which did not yet contain fissionable material. The building and equipment were damaged, but there were no injuries.

The Fessenheim plant would be the most powerful in France when the first two units of 890 megawatts each went into operation in February 1976.

Plan for Fessenheim A-plant dropped— Plans by the electricity authorities to build the Fessenheim nuclear power plant under construction near Strasbourg were dropped Nov. 20, 1976 at a meeting of the Franco-German regional committee that oversaw the area's development. The location of the planned station at Marckholsheim was one of 46 sites in France chosen for nuclear facilities.

Environmentalists had opposed French officials and industrialists over the issue and had demonstrated Oct. 10 in Strasbourg against the project.

Power station bombed. A nuclear power station in Brittany was damaged by two explosions Aug. 15, 1975. There was no danger of radioactivity from the damage, but power production was stopped. Although the explosions were deliberate, no group had claimed responsibility.

Several demonstrations had been staged in the area Aug. 15-17, protesting the installation of nuclear reactors.

Clash at A-plant site. Anti-nuclear demonstrators July 31, 1977 fought police during a massive protest march at France's largest nuclear reactor complex under construction at Creys-Malville.

At least 20,000 environmentalists gathered for the march. About 100 of them, wearing helmets and armed with clubs and fire bombs, left the main group of marchers and tried to break through the police line to reach the plant. The police fought back and the bulk of the marchers joined the fight or fled in a panic.

One demonstrator died, apparently from a heart attack, and about 100 persons were injured. These included several policemen whose hands were blown off when they improperly threw tear gas grenades.

The demonstrators included about 5,000 West Germans and an equal number of Swiss. A large number of Italians also were involved.

The more militant groups, in particular the 100 persons who began the fighting, were believed to have been West Germans. In reference to the West German presence, a police official said, "For the second time Morestel [a town near the plant] is occupied by the Germans." Anti-German sentiment had been a growing part of the French political debate before the Creys-Malville demonstration. (According to June 28 press reports, both left-wing and right-wing politicians had attempted to play on the traditional French fear of German power in recent speeches.)

While police blamed West German agitators for the violence, others sympathetic to the demonstrators charged the police with making improper arrests and showing no control during the emergency. No arrests were made during the fighting, and the 19 persons originally held by the police were caught after the demonstration in sweeps of the neighboring villages and campsites. The police reportedly vandalized cars with West German license plates.

Seven of the 19 persons arrested were released after questioning while 12 were held on various charges, mainly weapons possession. Seven of those charged were West Germans, two were Swiss, and three were French.

In addition to the controversy over the violence, the Creys-Malville protest brought the issue of nuclear development into the forefront of French political debate. President Valery Giscard d'Estaing had visited a reactor site at Pierrelatte July 30 to reassert his commitment to the nuclear program.

Interior Minister Christian Bonnet fully backed the police action at the protest, it was reported Aug. 2, and said the nuclear project was indispensable to the nation's energy independence.

The Communist Party endorsed the government policy of nuclear development and attacked the anti-nuclear demonstrators at Creys-Malville and elsewhere. However, the Socialist Party did not condemn the anti-nuclear forces. The Socialist-dominated labor federation had joined in several nuclear protests and had protested over police action at Creys-Malville.

The reactor at Creys-Malville was due to be completed between 1982–1983. It would be the first atomic power station in the world using plutonium in a fast breeder reactor, and would be the largest such facility in Europe. In full operation it would produce eight billion kilowatts per hour.

The project was a joint undertaking of four European nations, Italy, France, West Germany and the Netherlands. The decision to continue with its development had been made July 5 despite U.S. President Carter's request for a halt to development of breeder reactors and the use of plutonium fuel. The reactor was planned to cut down on the amount of uranium the four nations would need to import once their atomic plants were operational.

Uranium mines bombed—Five bombs exploded in the surface buildings of the French atomic authority's uranium mines at Margnac, it was reported Nov. 15, 197 . The explosions caused millions of dollars in damages to the structures and the extracting plant. The identity of the bombers was not discovered.

Margnac was located in the south-west corner of France. Its mine produced about 40% of the uranium obtained in the area.

GREAT BRITAIN

Windscale nuclear plant approved. The House of Commons voted overwhelmingly May 15, 1978 to provide funds needed to build a nuclear reprocessing plant at Windscale on the northwest coast of Great Britain. The plant was backed by the government and major opposition leaders and opposed by the Liberals.

The proposed Windscale plant had caused opposition from environmentalists who feared that the plutonium produced at the site could harm people in the area. A panel of inquiry headed by Justice Peter Parker reported to Parliament that the risk from the plant would be small.

The House of Commons included a provision in the approval of the plant that limited the amount of fuel to be reprocessed at the Windscale facility to 1,-200 tons a year.

According to press reports May 25, Japanese utilities already had signed $900 million in contracts with Great Britain for reprocessing spent nuclear fuels.

INDIA

Canada suspends A-aid to India. Canada suspended its nuclear aid to India May 22 because of India's detonation of a nuclear device May 18, 1974.

Canadian External Affairs Minister Mitchell W. Sharp had said May 22, 1974 that the Canadian government planned to review its other aid programs "to be sure that our priorities are the same as the Indians." He added that Canada was also asking other governments for joint consideration of "the broad international implications" of the explosion."

Sharp said, "What concerns us about this matter is that the Indians, notwithstanding their great economic difficulties, should have devoted tens or hundreds of millions of dollars to the creation of a nuclear device for a nuclear explosion."

The aid suspension would affect all shipments of nuclear equipment and material to India as well as technological information.

Canada had asked India to send high officials immediately for "a complete exchange of views" on the situation, Sharp said. He indicated that Canada particularly wanted to know the source of the plutonium used in the test.

News reports suggested that Canadian style nuclear reactors in India, built with Canadian aid, were more suited to the production of plutonium than were U.S.-designed reactors there.

A spokesman for the Canadian Ministry of External Affairs said May 28 that Canada was pressing "very actively" for international action to prevent any further spread of the capabilities to produce nuclear weapons. The statement was made following a news conference in Ottawa by visiting Pakistani Foreign Minister Aziz Ahmed, who called for international talks on "some way to prevent India from going ahead" with development of nuclear explosions.

Meanwhile, India defended itself against growing foreign criticism. Prime Minister Indira Gandhi said May 25 "allegations and apprehensions" that India was developing nuclear weapons were unfounded. In response to criticism that India was too poor a nation to spend resources on nuclear development, she said the same objection had been made "when we established our steel mills and machine-building plants." Such things were necessary, she said, because it was "only through the acquisition of higher technology" that poverty could be overcome. Indian newspapers denounced Western critics of the explosion.

Gandhi was reported to have sent a letter to Pakistani Prime Minister Zulfikar Ali Bhutto May 22, reiterating that India's explosion did not pose any threat to Pakistan's security. She was said to have reaffirmed India's commitment to peace agreements with Pakistan and to have denied that India had any ambition to "dominate or exercise any hegemony" over other subcontinent countries.

The Central Treaty Organization, at the end of a two-day ministerial conference in Washington, issued a statement May 22 that took note of India's nuclear test and merely expressed opposition to nuclear proliferation. The meeting was attended by the foreign ministers or other representatives of Pakistan, Britain, Iran and Turkey and U.S. Deputy Secretary of State Kenneth Rush, an observer.

In a related development, unidentified sources close to the Iranian government said Iran was negotiating with the U.S., the Soviet Union and Canada for nuclear technological assistance, according to a New York Times report May 21. As signs of Iran's intention to accelerate nuclear energy production, Iran had recently created an Atomic Energy Agency and a former president of Argentina's Atomic Energy Commission, Rear Adm. Oscar Armando Quihillalt, had assumed a role as an adviser on atomic energy, the report said.

U.S. halts uranium aid to India. The U.S. had suspended delivery of enriched uranium fuel to India until New Delhi pledged not to use the atomic fuel in any nuclear explosion, officials of the U.S. AEC disclosed Sept. 7, 1974.

Shipment was halted on an Indian order of enriched uranium for its atomic power plant near Bombay, built with U.S. assistance. Under a 1963 accord, the U.S. had promised to provide fuel for the reactor over 30 years.

The first of about four planned shipments under the order was made shortly after the Indian nuclear explosion in May, but further deliveries were stopped until New Delhi met the U.S. demands.

A spokesman for the AEC said Sept. 16 that India had agreed to give the specific assurances sought by the U.S. The issue had been discussed in Vienna, Austria in recent days by AEC Chairman Dixy Lee Ray and Homi N. Sethna, chairman of the Indian Atomic Energy Commission. The two were attending the annual conference of the International Atomic Energy Agency (IAEA), which met Sept. 16–20. (The IAEA general conference Sept. 16 admitted North Korea and Mauritius as new members of the agency, raising total membership to 107.)

The U.S. NRC July 21, 1976 approved the sale of 13.5 tons of uranium fuel to India, leaving unanswered the question of future exports to India.

U.S. may have aided Indian nuclear blast—U.S. Secretary of State Henry Kissinger acknowledged Aug. 2, 1976 that there was "a high probability" that heavy water supplied by the U.S. had been used by India in its nuclear explosion in 1974. The admission was contained in a letter to Chairman Abraham Ribicoff (D, Conn.) of the Senate Government Operations Committee, who released the letter Aug. 8. Kissinger said a "misinterpretation" of Indian assurances and of technical data had led the State Department in June to conclude incorrectly that no U.S. material had been involved in the Indian test. The State Department, according to Kissinger, now believed that some of the U.S. heavy water supplied to India had been present in a Canadian-supplied reactor during the period when India was amassing the plutonium required for its nuclear test.

In disclosing Kissinger's letter, Ribicoff said he was disturbed "because it indicated India has misused our peaceful nuclear assistance to develop its version of the atom bomb."

Canada's India A-aid voided. The Department of Industry, Trade and Commerce officially cancelled an export permit covering an estimated $1.5 million in nuclear-related material and equipment sales destined for use in a heavy-water atomic reactor in India, the Toronto Globe and Mail reported Jan. 18, 1975. Ottawa had suspended all such sales in May 1974 after India exploded a nuclear device.

Carter approves uranium sale. U.S. President Carter April 27, 1978 approved the sale of U.S. enriched uranium to India for its nuclear power plant at Tarapur. His action overrode a decision April 20 by the Nuclear Regulatory Agency to bar the sale on the ground that it would not meet the requirements of the Nuclear Nonproliferation Act.

In a message to Congress, Carter said refusal to export the uranium to New Delhi "would seriously undermine our efforts to persuade India to accept full-scope safeguards and would seriously prejudice the achievement of other U.S. nonproliferation goals." India, the President added, had assured the U.S. that it would "use our exports only at the Tarapur atomic power station and not for any explosive or military purpose. . . ."

Prime Minister Morarji R. Desai had told Parliament April 23 that India would seek alternative fuel to avoid shutting down the Tarapur plant. In another statement to Parliament April 24, the prime minister had warned that if the U.S. Nuclear Regulatory Agency's decision to block the sale were not reversed, it would be a violation of a 1966 U.S.-Indian agreement and would free his government "to adopt any course we like to safeguard our interest." He said he viewed the agency's action "with considerable disquiet and disbelief."

The House of Representatives rejected a resolution July 12 that would have barred the sale of seven tons of uranium to India.

The vote was 227–181. It was a victory for President Carter, who supported the sale. Rep. Jonathan Bingham (D, N.Y.), a supporter of the sale, said that Carter was "convinced that to deny the license for this shipment would certainly impede further negotiations to achieve greater cooperation in nuclear matters."

Opponents of the sale argued that the U.S. should refuse to provide uranium to India because India did not accept international safeguards on its nuclear facilities. Rep. Richard Ottinger (D, N.Y.), a sponsor of the resolution to block the sale, said that the "whole world is watching to see whether the U.S. Congress takes seriously the Nonproliferation Act it approved earlier this year."

Ottinger continued, "If we do not apply the act with respect to India, we have no reason to expect anybody to think we would apply it to anybody else."

But supporters of the sale maintained that denial of the uranium would only hurt relations between the two countries, while India would still be able to acquire the uranium from some other country.

The House action meant that the sale would go through. Under the new anti-proliferation law, Congress had 60 days after the president's approval of a sale to block the export by a concurrent action. The Senate did not vote on the sale.

Loss of U.S. A-device disclosed. Prime Minister Morarji R. Desai April 17, 1978 confirmed a report published in the U.S. that an American nuclear powered spy device had been lost in the Himalayas in 1965.

Desai's announcement to Parliament, however, disputed an assertion in an article by Outside (a publication of the magazine Rolling Stone) that the Indian government had not been consulted on the operation at the time. He said that the mission to plant the device, which was designed to monitor nuclear explosions and missile tests in neighboring China, had been undertaken after joint consultation between the U.S. and India "at the highest level."

Desai made the disclosure following protests in India that radioactive material from the lost device could contaminate the Ganges River. He said that tests conducted in the past few days with the cooperation of the U.S. government indicated that there "was no cause for alarm on grounds of health or environmental hazards." Desai said the device weighed 38 pounds, had "a power system energized by two or three pounds of plutonium-238 metal alloy, contained in several doubly capsulated leaktight capsules." It had been lost during a blizzard as a U.S. expeditionary team was attempting to place it on a Himalaya mountain top, the prime minister said. Ground and air searches

during the following three years, he said, failed to find any trace of the instrument.

The article by Outside said a Central Intelligence Agency mountain team had tried to place the nuclear powered device atop Nanda Devi, a 24,645-foot peak. Bad weather stopped the climbers 2,000 feet from the top, and the powerpack was left in some rocks only to be buried by an avalanche by the time the climbers returned a year later, according to Outside.

The U.S. State Department refused April 17 to confirm or deny Desai's statement. The CIA refused comment.

ISRAEL

Uranium 'loss' said Israeli plot. The apparent disappearance of 200 metric tons of uranium ore on the high seas in November 1968 was the culmination of a carefully orchestrated plot by Israeli intelligence agents, Time magazine reported May 30, 1977. The incident appeared to support that public safety was endangered in many countries by lax controls over atomic fuels.

Time said that after several weeks of investigation by a team of correspondents, it had learned that the plot had been devised to disguise a secret purchase of uranium by Israel for its nuclear reactor at Dimona in the Negev Desert. Secrecy was necessary, Time said, because an open purchase might have caused the Soviet Union to supply nuclear arms to the Arab nations.

According to Time, Israel received assurances from the West German government of Chancellor Kurt Georg Kiesinger that it would be permitted to disguise the purchase of uranium as a private commercial transaction. In return, Time said, Israel promised Germany access to its advanced uranium separation process that could be used to produce nuclear weapons. (West German officials now refuse either to confirm or deny government involvement in such a deal, Time reported.)

Some $3.7 million worth of uranium ore was purchased, in what later became Zaire, from the Belgian firm Societe General des Minerais, according to the magazine. It was bought by a now-defunct German petrochemical company, Asmara Chemie, which Time said had never before bought uranium. The two companies told the European Community's atomic energy agency, Euratom, in August 1968 that the ore would be shipped to SAICA, a paint company in Milan, which Time said also had never been known to use uranium. (The destination had been shifted from the original choice, a Moroccan firm, because Morocco was not in the European Community and nuclear material could not be shipped there without a special permit.)

SAICA was sent $12,000 to buy equipment to mix the uranium and paint, Time said. But a few days after the uranium was shipped, Asmara reportedly told SAICA the ore ship had been lost and to keep the money.

Time sources said the uranium was shipped from Antwerp, Belgium aboard the *Scheersberg A,* a tramp steamer secretly owned by the Israeli intelligence agency, Mossad. The uranium was transferred, under cover of darkness, to an Israeli ship in the Mediterranean Sea between Cyprus and Iskenderun, Turkey. The Israeli boat then took the uranium to Haifa, according to Time. Port records showed that the *Scheersberg A* arrived empty in Iskenderun Dec. 2.

The *Scheersberg A,* according to Time, "was almost certainly involved" in the refueling of five French-built gunboats seized by Israeli agents from Cherbourg harbor in December 1969. *The Scheersberg A* had since been sold and renamed twice and currently was making a regular run between Greece and Libya under the name *Kerkyra.*

A Euratom investigation of the missing uranium had been hampered by the agency's lack of police powers, Time said, and after a few months Euratom had asked for help from the security forces of Western nations. A West German investigation, Time said, had been "abruptly—and mysteriously—halted" soon after it began in 1969.

JAPAN

Nuclear ship springs a leak. Japan's first nuclear-powered ship, the 8,214-ton Mutsu, was finally permitted to return to its home port of Mutsu at the northern tip of Honshu Oct. 15, 1974 after drifting off the coast for nearly six weeks.

The ship, whose reactor had sprung a leak during a test run in the North Pacific, had been barred from returning by protesting fishermen who feared that the vessel would contaminate their scallop beds in Mutsu Bay. The fishermen had begun their protest action before the Mutsu sailed Aug. 25, setting up a line of boats in the bay to prevent its departure; the vessel managed to slip out to sea during a storm. The government had ordered the ship to return to port for repairs after the reactor began leaking Sept. 1, but the fishermen refused to permit this. Emergency repairs were made at sea.

Forty-four of the 55-man crew had abandoned the ship Oct. 7, leaving only a skeleton force aboard.

NORWAY

Study Backs Nuclear Power. Nuclear power was an acceptable energy option for Norway, according to the majority report of a 21-member study committee. After a two-year study, the committee's findings were submitted to the Ministry of Petroleum and Energy Oct. 12, 1978.

The majority report, backed by 18 members of the committee, made its approval of nuclear power conditional on the finding of a satisfactory solution to the problem of disposing of the radioactive wastes produced by nuclear facilities. If a means of handling the wastes could be found, the danger of accidents at a nuclear plant was not great enough to warrant rejecting nuclear power, the majority concluded.

Three of the committee members contended that nuclear power posed too great risks to be acceptable.

SOVIET UNION

Soviet A-accident. Travelers to Moscow said Feb. 20, 1970 that a blast had occurred in the Soviet Union's main nuclear submarine plant near Gorky, 250 miles east of Moscow. Several workers were said to have been killed and the Volga river polluted with radioactive waste. According to the Washington Post Feb. 25, Soviet sources said that radioactive material had leaked from a container in a shed at the plant, killing two employees and injuring three others. The accident, according to the Soviet sources, had occured in January, had been confined to the shed, and had not polluted the Volga.

SWITZERLAND

Marchers demand nuclear ban. Over 5,-000 protestors marched through northern Switzerland May 28–29, 1977 to demand a four-year ban on the construction of atomic power stations. The march began in Kaiseraugst, where 1975 protests had halted a planned power station, and it ended 30 miles away in Goesgen, site of another planned station.

In a related action June 12, voters in Basel overwhelmingly approved a local law that required city officials to oppose the construction of nuclear power plants. The vote was the first nuclear power referendum in Western Europe, according to local officials. The federal government had planned a complex of nuclear plants in the area.

WEST GERMANY

230 injured in nuclear plant protest. A demonstration by 30,000 persons Nov. 13, 1976 at the Brokdorf nuclear power plant building site, 25 miles northwest of Hamburg on the Elbe River, erupted into violence when about 3,000 of the more militant protesters attempted to storm the enclosed area. Police used fire hoses, tear gas and clubs to repel the assault. At least 151 demonstrators and 79 policemen were hurt.

The protest underlined a growing popular resistance to the West German nuclear-power program, which had begun in 1974 as an answer to the oil crisis. The government had projected a nuclear electrical capacity of 45,000-50,000 megawatts (mw) by 1985. Local protests and increased safety requirements had cut the government's projections by 10,000 mw, it was reported Nov. 11.

Current megawatt capacity was 6,400, with about 12,000 under construction and an additional 9,000 planned.

Opinion polls, made public Nov. 11, indicated that 20% of the West German population was concerned over expansion of nuclear-energy facilities. The opposition in Bonn had accused the government of planning the growth of the nuclear energy industry, then backing away from specific projects after those projects were criticized, it was reported Nov. 11.

Other nuclear protests had been centered in Lower Saxony, where feelings were reportedly more intense. The government planned to build West Germany's first full-scale atomic fuel reprocessing and disposal plant in Lower Saxony. The area had large underground salt strata where the government hoped to bury radioactive waste from all West German nuclear power stations. Lower Saxony residents were incensed by the plan and had protested repeatedly.

Earlier in the year Interior Minister Werner Maihofer had announced that no new nuclear-plant construction permits would be issued until plans existed for the disposal of spent fuel. Three Bonn government ministers visited the premier of Lower Saxony Nov. 11 to pressure him into reaffirming his support for the disposal project.

Ban on A-plants, safety & waste disposal problems—West Germany's most industrialized and populated state, North Rhine-Westphalia, reported Jan. 13, 1977 that it would not build more nuclear plants until the problem of spent atomic fuel was settled. Heinz Kuhn, premier of the state, said there was no chance that plants already in operation would be shut down. Four of West Germany's 13 operating atomic-power plants were located in North Rhine-Westphalia.

Public protests in West Germany over the disposal of radioactive waste and worry over the safety of the plants themselves had led the federal government in 1976 to cut back the number of power stations planned for the country to 25 from 35.

The violence that had accompanied the November 1976 anti-atomic power demonstration in Brokdorf on the Elbe River had a special impact on Bonn politicians and had led to a debate on the value of nuclear development itself. Hans Matthofer, technology minister, said the Brokdorf protest represented "not just resistance to a nuclear power plant. The Elbe is a polluted river," Matthofer said, " . . . and people are protesting against all of this. Nuclear energy is just the symbol."

Public concern over the safety of the atomic plants was increased Jan. 14 when the ministry for protection of the environment in the southern state of Bavaria shut down an atomic-power station near Gundremminger after radioactive steam escaped. No one was injured or endangered by the leak, but the plant was closed indefinitely.

It was the third incident at the Gundremminger plant. Two men had been killed a year earlier by hot steam while working on a burst valve, and the plant had been closed three days in December 1976 because of a leak in the circulation system.

The events leading to the Gundremminger closing were similar to circumstances that led to the shutdown of the power station in Biblis, which had been a showpiece of the West German nuclear industry. The plant was closed for months in 1976 because of a problem in its main coolant circuit.

In addition to plant safety, the problem of fuel disposal had become more serious since the Brokdorf demonstration. Government experts had said that a central underground dump would be needed within 10 or 20 years to handle the mounting waste from the reactors currently in operation. Lower Saxony, the only West German state with the salt caverns necessary for disposal, had resisted attempts to locate the dump within its border. The governor of the state, Ernst Albrecht, told the federal government that atomic waste should be shipped to the U.S. rather than to his state.

The limits on domestic development put pressure on the nuclear power industry to increase exports in order to make up for the closed domestic market. To fully employ the nation's nuclear work force, eight power stations had to be produced each year.

Other Developments

5 nations maritime transport pact in force. The international convention on civil liability in the field of maritime transport of nuclear materials entered into force July 15, 1975 after five nations had ratified the pact. The five nations were France, Spain, Denmark, Sweden and Norway.

The convention had been adopted in December 1971 by an international conference in Brussels, held under the joint auspices of the Intergovernmental Maritime Consultative Organization, the Nuclear Energy Agency of the Organization for Economic Cooperation and Development and the International Atomic Energy Agency.

The new convention would free shipowners from liability under international maritime law in the case of nuclear damage involving the transport of nuclear materials to or from a nuclear installation. The nuclear installation would be liable for all damages.

A-Test Controversies & Counter-Measures

Peaceful Tests Bring Problems

First U.S. 'peace' test. The U.S.' first atoms-for-peace test was carried out Dec. 10, 1961 with the detonation of a five-kiloton nuclear device 1,216 feet underground at a test site 25 miles southeast of Carlsbad, N.M.

The test—dubbed Project Gnome—was not an unqualified success. The nuclear device, exploded in an 8-by-10-foot chamber cut into a strata of rock salt, produced pressures greater than expected. It rocked the surrounding desert floor and caused clouds of intensely radioactive steam to escape from the mouth of a vertical shaft connected with the detonation chamber by a 1,000-foot tunnel. Radioactivity reached 10,000 roentgens an hour (maximum safe exposure: 1/2 roentgen yearly) at the shaft's mouth. The Atomic Energy Commission closed nearby highways and took other local precautions. AEC officials said that residents of Carlsbad and other nearby communities were in no danger.

Gnome was the first explosion planned under the AEC's Project Plowshare program for the development of nuclear explosions for peaceful purposes.

Gnome was designed for the following peaceful applications: (a) to test whether heat from an underground explosion could be stored and tapped for steam to power electrical generators (a 750-foot shaft was sunk into the desert floor over the detonation chamber to see whether, when filled with water, it would yield usable steam); (b) the production of gaseous and solid radioactive isotopes; (c) to measure the characteristics of nuclear detonations in salt formations; (d) to obtain neutrons produced by the detonations for laboratory analysis; (e) to test the design of nuclear devices intended for peaceful detonations.

Fallout & Counteraction

Anti-radioactive fallout program begins. The first statewide program to reduce the quantity of fallout-caused radioactive substances in milk was begun in Minnesota Aug. 23, 1962. An estimated 40% of Minnesota's farmers complied with state requests to voluntarily withdraw their dairy cattle from pasture and begin feeding them hay and other dried feed that had been aged at least 21 days in order to reduce the relatively high levels of radioactive iodine-131 reported in Minnesota milk samples. The substance produced by atmospheric nuclear bomb tests and borne to the northern U.S. by fallout-contaminated winds, was known to concentrate in the thyroid glands of children and to pose the threat of cancer. The eight-day halflife (period in which the substance was reduced by half through radioactive decomposition) of iodine-131 made its curtailment in fresh milk feasible if cattle were kept from pasture, where the substance was likely to have been newly deposited by rain and wind, and instead were given aged feed from which it had dissipated.

The Minnesota program was described as precautionary, but Dr. Warren R. Lawson, head of the Minnesota Health Department's Radiation Section, warned Aug. 23 that "if we get any large amount of Russian fallout" from the U.S.S.R.'s new test series, it might be necessary to obtain 80% farmer compliance to hold iodine-131 levels below the maximum permissible dosage set by the Federal Radiation Council. The Council's recommended maximum was 100 micromicrocuries of iodine-131 per liter of milk per day: current Minnesota readings averaged 46, and some milk sampling stations reporting readings of over 100. The New York Times reported Aug. 24 that in the Twin Cities area, the dosage had averaged nearly 100 mmc. per liter per day for the past 11 months.

The U.S.' first official estimate of the number of persons who might be adversely affected by radiation from atmospheric nuclear tests had been made by the FRC in a report released June 1. The report said that the radioactivity caused by fallout from tests thus far was only a fraction (5%) of that resulting from natural "background" radiation in the earth and its atmosphere. It estimated that over the next 70 years a total of 2,700 persons might die prematurely from leukemia and bone cancer induced by radiation from fallout: this compared with medical estimates that 980,000 persons would suffer the same diseases from other sources during the same period. It predicted that test-caused radiation might cause 110 grossly abnormal births throughout the world in the next generation; four million to six million were expected from natural causes. It concluded, however, that "any further increase in testing will, of course, increase the [radiation] exposure" and, presumably, the threat to human health.

Dr. Russell H. Morgan, chairman of the Public Health Service (PHS) committee

and chief radiologist of Johns Hopkins Hospital (Baltimore, Md.), testified before a joint Congressional Atomic Energy subcommittee June 7 that continued Soviet atmospheric testing might make it necessary to remove contaminated milk from the diets of children and pregnant women. He warned that an increase in fallout probably would cause the quantities of iodine-131 in fresh milk to approach or exceed the guidelines established by the FRC. He said there was "accumulating evidence that radiation delivered to the neck and throat of infants and children may induce cancer of the thyroid gland after the elapse of a number of years."

The FRC's recommended permissable maximum dosage levels for the human thyroid gland was 1.5 rem annually for the individual, with the average dosage not to exceed .5 rem. According to the June 8 New York Times, the PHS had estimated the average U.S. child's dosage as .16 rem in Sept. 1961–Jan. 1962 but had estimated that 4% of these dosages had attained .4 rem. Recent dosages to children in some areas of Minnesota and Iowa were said to have approached the permissable maximums.

Criticism of the U.S.' fallout monitoring system was amplified June 5 by Irving Michelson of the Consumers Union, a private group studying radioactivity in foods with AEC and PHS support. Michelson told the Congressional subcommittee that the PHS' air sampling network was "of very limited value, if not misleading," and that information about radiation levels was not "available soon enough to take any protective action." He said that, under current procedures, total diet samples "do not give us data until three or four months after the food has been eaten" and that the PHS' milk monitoring service required two to three months to analyze the strontium-90 content of fresh milk. He said this delay had prevented the PHS from immediately detecting the high amounts of iodine-131 present in U.S. milk supplies after the USSR's resumption of atmospheric tests in 1961.

Fallout in 1962—Reports of radioactive fallout in the U.S. during 1962:

PHS officials reported in Washington Apr. 12 that a sharp but undefined rise in radioactive substances from fallout had begun in Southern U.S. milk supplies. The rise was attributed to early pasturing of cattle in Southern states. The Naval Research Laboratory in Washington reported the same day, however, that gross radiation from fallout had not risen as sharply in March as in previous years despite the resumption of Soviet tests in the atmosphere.

The New York Times reported May 6 that PHS officials had comfirmed a rise in iodine-131 in certain U.S. cities' food supplies in late 1961 to "range III" levels requiring countermeasures if sustained. No such measures were taken, nor were any envisaged, it was reported, because the levels did not persist more than a brief time in any one place.

The New York City Department of Air Pollution Control reported May 13 that radioactive contamination of the city's air had increased by 650% and radioactive sootfall by 400% during 1961. New York officials did not consider the current levels dangerous.

A Food & Drug Administration survey released May 27 reported that an analysis of commercial baby foods showed them to contain no more than 4% of the amount of strontium-90 considered a potential threat to infants' health. It said that no "change in normal dietary patterns" was warranted.

A sharp increase in the amounts of fallout-caused iodine-131 in fresh milk was reported by the PHS in figures issued May 23 and 30. The May 30 figures, based on readings taken May 14–18, showed the following levels (in terms of micro-microcuries of iodine-131 per liter of milk): Wichita, Kan. 660 mmc.; Kansas City, Mo. 600 mmc.; Des Moines, Ia. 330 mmc.; Minneapolis 290 mmc.; Chicago 90 mmc.; St. Louis 80 mmc.; Cincinnati 50 mmc.; Denver 45 mmc.

Surgeon Gen. Terry said May 30 that the milk surveillance program had been increased but that "the readings do not call for protective measures of any sort."

Declines in the high Midwestern readings were reported by the PHS in figures made public June 21. The Midwestern fallout was attributed to the U.S. atmospheric nuclear tests in the Pacific. The readings had remained constant through June 14 but then showed the following reductions reported June 21: Wichita 80 mmc.; Kansas City, Mo. 155 mmc.; Des Moines 70 mmc.; Minneapolis 30 mmc.; Chicago 10 mmc.; St. Louis 35 mmc.; Cincinnati 10 mmc.

Two groups of children—10 each from Kansas City, Mo. and from St. Louis— were examined by the NYU Medical Center and were found to have absorbed only slight quantities of iodine-131 in their thyroid glands despite the relatively high levels reported in the milk they drank. The Kansas City group was tested June 22, the St. Louis group June 27. Dr. Merril Eisenbud of NYU reported to the PHS July 7 that the amounts of iodine-131 detected had not approached the recommended level for countermeasures.

Dr. G. D. Carlyle Thompson, Utah health director, reported Aug. 1 that milk from certain parts of northeastern Utah was being diverted from Salt Lake City because of increases in iodine-131 as a result of an atmospheric test carried out July 7 at the AEC's Nevada proving grounds. The level of radioactive iodine in milk supplied by the affected areas was not detectable July 6 but had risen to 1,660 mmc. a liter by July 20. It fell to 450 mmc. during the next few days, but rose again to 2,050 mmc. July 25.

The PHS reported Aug. 17 that the July 7 Nevada test had been responsible for an increase in the quantity of iodine-131 to 580 mmc. per liter of milk in the Salt Lake City area and 370 mmc. in Laramie, Wyo. milk supplies. Both figures were averages for the entire month of July. The July average for the entire PHS sampling network (61 stations throughout the U.S.) was 40 mmc. of iodine-131 per liter, compared with 30 mmc. in June.

Fallout peril denied. The Kennedy Administration made public Sept. 17, 1962 a report asserting that existing federal guidelines for atomic radiation were inapplicable to radiation received from test-caused fallout and that no fallout in the U.S. had reached dangerous levels.

The report, written by the Federal Radiation Council and issued by HEW Secy. Anthony J. Celebrezze, council chairman, declared that radiation levels, even if they exceeded the FRC's recommended maximums, "would not result in a detectable increase in the incidence of disease." The FRC said that its radiation recommendations had been written to govern possible dosages received from peaceful nuclear projects and medical usage. It said that these recommendations deliberately had been set far below radiation levels known to be harmful and that they could not be applied with meaning to cases in which radiation dosage had been received from fallout or other uncontrolled sources.

The report criticized Utah and Minnesota health officials for measures intended to reduce the amount of radioactive iodine-131 from fallout in their milk supplies. Iodine-131 levels in Minnesota and Utah milk had reached or exceeded the FRC's recommended maximum levels. The FRC statement said that no such measures were needed yet to protect consumers from fallout residues in their food. It warned that the "counter-measures may have a net adverse rather than favorable effect."

The 2d report of the U.N. Scientific Committee on the Effects of Atomic Radiation, issued at UN headquarters Sept. 9, contained the warning that fallout from continued nuclear testing might eventually affect man's genetic heritage. The report accepted U.S. estimates that test-caused radiation did not yet exceed a fraction of natural radiation levels and did not constitute a specific health hazard. But, with respect to possible genetic effects of fallout, it declared: "Because of the available evidence that genetic damage occurs at the lowest levels experimentally tested, it is prudent to assume that some genetic damage may follow any dose of radiation, however small." It acknowledged that scientists "still know very little" about the harmful effects of low-level radiation, but it warned: "There should be no misunderstanding about the reality of genetic damage from radiation. Although individual mutations vary greatly in their effect, there is no doubt that any increase in mutation is harmful." The report was the first to be issued by the 15-nation U.N. committee since its initial findings were published in 1958.

The PHS said Dec. 6 that radioactive iodine-131 fallout was still at a "warning level" for Palmer, Alaska, although fallout was waning over most of the U.S. The PHS had reported Nov. 14 that iodine-131 levels had exceeded peacetime protective standards for Palmer as well as Salt Lake City, Utah during the previous 12 months. The yearly total approached permissible levels in several Midwestern sampling stations but was otherwise low in the U.S.

(These were the first PHS-published reports to give 12-month figures for iodine-131, strontium-89 and strontium-90 in different areas of the U.S.)

The Joint Congressional Committee on Atomic Energy had charged the executive branch Sept. 26 with failing to develop adequate protective standards or organization on fallout problems.

Fallout rise reported. The Federal Radiation Council reported May 31, 1963 that U.S. and Soviet nuclear tests carried out in 1962 had doubled the amount of radioactive debris in the earth's atmosphere and had caused a commensurate rise in the intensity of test-caused fallout and radioactivity in human food supplies. The Council asserted, however, that even with the increased fallout, "the health risks from radioactivity in foods ... are too small to justify countermeasures."

According to the Council's report, the average measurement of radioactive strontium-90 in the U.S. diet rose from a range of 4 to 8 Sr units in 1961 to a range of 8 to 13 Sr units in 1962 and probably would rise to approximately 50 Sr units in 1963. (One Sr unit meant the presence of one trillionth of a curie of radiation per gram in calcium contained in the matter under analysis.) The new level of radiation exposure from fallout was expected to amount to 1/30 the exposure from natural radiation over a 30-year period (the human reproductive span) or 1/20 the natural dosage over a 70-year period (the human life-span). It was expected to double the estimated risk of fallout-caused genetic damage to newborn infants from one in 1,000,000 births to one in 500,000.

The report disclosed that in 1962 Soviet atmospheric tests had yielded 180 megatons of energy and U.S. tests had yielded 37 megatons of energy. Sixty megatons of the Russian yield was described as highly-radioactive fission energy, and 16 megatons of the U.S. tests' yield was in the same category. (The AEC had reported 39 Soviet atmospheric tests and 36 U.S. atmospheric tests during 1962; Joseph Lee Porter, a member of the Space Science Board of the U.S.' National Academy of Science, disclosed at the Committee on Space Research meeting in Warsaw June 3 that there had been additional unannounced Soviet high-altitude tests in 1962.)

FRC Executive Director Paul C. Tompkins testified before a subcommittee of the Joint Congressional Atomic Energy Committee June 3 that the Council believed it was necessary to establish permissible maximum fallout radiation standards for the general public. Tompkins declared that the Council's radiation guidelines for industrial radiation had not been meant to apply to the general public and were not a meaningful measure of the radiation danger from fallout. The FRC guidelines for workers subject to industrial radiation were: Range I—0–20 Sr units (no action); Range II—20–200 Sr units (government surveillance instituted); Range III—200–2,000 Sr units (radiation counter-measures considered and applied if necessary).

The FRC's industrial radiation guidelines had recommended maximum permissible dosages of 150 Sr units for the general population, if exposed to such radiation, and of 50 Sr units for newborn infants suffering such exposure. It was these recommended maximum dosages that had aroused Congressional concern at the fallout increase and that the FRC denied could be applied to the fallout problem. Tompkins suggested that a more meaningful limit for the general public's exposure to fallout might be the maximum 2,000 Sr unit-dosage permitted atomic workers.

Rep. Melvin Price (D, Ill.), chairman of the subcommittee, charged June 3 that Tompkins' proposal for greatly increasing the public radiation level officially tolerated was "boosting the limits so the risks do not look like hazards." Price charged the Administration with failing to inform the general public of the best scientific estimates of how much fallout radiation might be safely tolerated.

Other Problems & Disputes

A-test blamed for death. Patrick Stout, 55, a former army sergeant who entered an atomic bomb crater at Alamogordo, N.M. in 1945 to show it was harmless, died of leukemia April 18, 1969. Doctors suspected his death had been caused by atomic radiation in the crater.

According to Stout's wife, he had worn only canvas coveralls and canvas boots when he entered the crater. A special panel of doctors and scientists had recommended that Stout be given service disability benefits when he was discharged from the army in 1955. He had contracted leukemia in May 1967.

Symposium debates radioactive fallout. Controversy over the effect of radiation on the environment marked a three-day symposium on engineering with nuclear explosives Jan. 14–17, 1970. AEC member Dr. Theos J. Thompson, said at the opening of the sessions, sponsored by the American Nuclear Society in cooperation with the AEC, that concern by environmentalists about radiation hazards was reaching almost ridiculous levels. He added that such persons did not evaluate the benefits of programs to develop peaceful uses of atomic energy, but had concentrated on very low level effects "whose extreme extrapolation might be detrimental."

Rep. Chet Holifield (D, Calif.), chairman of the Joint Committee on Atomic Energy, criticized the Administration's decision to cut funds for Project Plowshare, a 13-year program by the AEC to find peaceful uses for atomic explosions, from $29 million to $14 million in 1969. Holifield predicted more cuts in the program in the 1971 fiscal budget.

Dr. Edward L. Teller, a leading developer of the hydrogen bomb, said Jan. 15 that radiation from peaceful uses of atomic energy "can be easily guarded against." Teller said of the AEC's programs that "no big-scale enterprise has ever been carried out with more assurance [of public safety] than the atomic energy enterprise."

Robert B. Miller, an attorney for the American Civil Liberties Union, charged in a federal district court in Denver Jan. 12 that safety standards of the AEC "grossly underestimate" potential damage from radiation. Miller made the charge at the reopening of a suit to prevent natural gas from being freed from an explosion conducted Sept. 10, 1969—Project Rulison—in Colorado. He also asserted that the AEC had violated its own standards by conducting the test when wind directions had not been correct.

A-test site contaminated. In a report made public Aug. 20, 1970 the AEC disclosed that a 250-square-mile area of its Nevada test site was contaminated with poisonous and radioactive plutonium 239.

The report, submitted in July to the White House Council on Environmental Quality, said the contamination had occurred as a result of tests conducted in 1958 to determine whether the crash of a U.S. bomber could trigger a nuclear explosion.

Large parts of the contaminated zone, which constituted one-fifth of the AEC's Nevada test area, were reported Aug. 20 to have been sealed off to the public. An AEC spokesman said, however, that health hazards could result only if the plutonium was stirred up into the atmosphere by a wind to pollute the air.

The report revealed that one out of every 12 underground atomic tests carried out since the 1963 treaty banning atmospheric tests had leaked some radioactivity "off site," but the amount was said to be one-tenth of the public exposure level considered safe by the Federal Radiation Council.

A-test cloud vented. The AEC announced it had conducted two nuclear explosions Dec. 16, 1970 and one each Dec. 17 and Dec. 18 at its Nevada test site.

A cloud of radioactive dust was blown into the air after the Dec. 18 test, resulting in the evacuation of some 600 workers from the area. All underwent "standard decontamination procedures," according to an AEC spokesman, who described the amount of radioactivity offsite as "very low."

By Dec. 19 the cloud had traveled some 450 miles away from the site to central Utah. AEC officials refused to speculate whether the cloud would remain intact long enough to cross a national boundary and thereby violate the 1963 treaty limiting nuclear tests. Tests in the contaminated area were suspended Dec. 21 pending a radiological survey to assure the safety of AEC workers. The radioactive cloud was reported that day to have disappeared over Wyoming.

A-tests resume—Officials of the AEC said May 14, 1971 that underground tests at the Nevada tests site would resume "in early June," following suspension of all tests during investigation of the circumstances that caused a test Dec. 18, 1970 to give off a cloud of radioactive material.

The AEC report said the venting of radioactivity in the Dec. 18 test, codenamed Baneberry, had been caused "primarily by the earth around the explosive device being more saturated with water than had been anticipated." Test resumption was approved "under more stringent and detailed technical analysis and review procedures," including "a closer examination of the geology of test locations."

In a related development, the Swedish government delivered a memorandum April 30 to the U.S. embassy in Stockholm declaring that an increase in radioactivity over Sweden in December 1968 could be attributed to a U.S. underground nuclear test Dec. 8, 1968. The note said the test was in violation of the 1963 treaty limiting the release of test radioactivity across national borders. An AEC spokesman replied April 30 that of some 30 underground tests conducted by the U.S. in 1968 only one had caused a "very small amount of radiation" offsite.

Court battle to ban Alaskan A-test. The most powerful underground nuclear test conducted by the U.S. was executed Nov. 6, 1971 on the Alaskan island of Amchitka in the Aleutians, five hours after the Supreme Court denied a last-minute appeal by environmental and antiwar groups for a temporary injunction.

The test, code-named Cannikin, was conducted in the face of strong opposition from Canada, Japan, some U.S. senators and thousands of U.S. citizens, all of whom feared possible adverse ecological consequences. However, the earthquake and tidal wave feared by test opponents failed to materialize and the AEC reported the absence of any trace of radioactive leakage.

The underground cavern created by the blast collapsed Nov. 8, forming a large crater on the island's surface. The collapse registered 5 on the Richter scale.

First reports released Nov. 8 on the test's environmental effects indicated that several rockfalls, "larger than expected," had destroyed four eagle nests and a peregrine falcon nest and killed 13 birds. No falcons or eagles, considered two of America's rarest birds, were killed.

Supreme Court bars delay—Meeting in special session to hear oral arguments for and against the test, the Supreme Court Nov. 6 voted 4–3 against a temporary injunction. A last-minute brief filed Nov. 4 by the Committee for Nuclear responsibility Inc. and seven other environmental, antiwar and American Indian groups had asked for at least a week's postponement of the test to hear their contention that the AEC had downplayed the test's environmental risks in violation of the 1969 National Environmental Act.

Voting to deny the injunction were Chief Justice Warren E. Burger and Justices Harry A. Blackmun, Potter Stewart and Byron R. White.

Dissenting from the majority opinion, Justices William J. Brennan Jr., William O. Douglas and Thurgood Marshall advocated a delay to enable the court to decide whether to review charges that the AEC had violated the environmental law's requirement for a statement on the test's environmental impact that included "responsible opposing views." Supreme Court Justices Brennan and Marshall thought there was a "substantial question" of such a violation. Justice Douglas noted what he called "obvious defects" in the impact statement.

Solicitor General Erwin N. Griswold argued the test was vital to national security and said the question of whether to conduct the explosion should be "left to the determination of the politically responsible officers of the government," not to the judiciary. He warned that a delay might cause complications and lead to a year's postponement of the test.

David L. Sive, a New York lawyer representing the Committee for Nuclear Responsibility, argued that a lower court of appeals had already ruled that the AEC might have violated the law. He said the omission of "responsible opposing views" from the environmental impact statement made the statement "substantially misleading" and therefore invalid. He urged the temporary delay to permit the court to hear the merits of his case.

The Supreme Court decision climaxed a four-month legal battle to block the test. The events leading to the decision:

U.S. District Court Judge George L. Hart Jr. Aug. 26 dismissed a suit seeking release of secret government papers that had allegedly advised against the test. He subsequently ruled Aug. 30 that the Cannikin test would comply "with all relevant laws and treaties" and refused to delay or cancel the blast.

Ordered by the Court of Appeals for the District of Columbia to reopen the case, Judge Hart Oct. 26 directed the government to furnish him with secret documents concerning the test's environmental impact so he could determine whether to release them to the environmental groups. He stayed his order pending review by the court of appeals.

The court of appeals Oct. 28 upheld Judge Hart's request for the documents. It also refused to block the test temporarily, contending that such an action "would interject the court into national security matters that lie outside its province."

After examining the documents, Judge Hart Nov. 1 refused for the second time to grant a preliminary injunction against the test. He said the Committee for Nuclear Responsibility had failed to prove that the AEC had withheld information on possible environmental hazards or had failed to consider the hazards in its decision to hold the test.

In compliance with Judge Hart's ruling, the AEC Nov. 3 released 187 pages of secret documents. One was a secret memorandum dated Dec. 2, 1970 from Dr. Russell E. Train, chairman of the President's Council on Environmental Quality, to Undersecretary of State John N. Irwin 2nd. The memo had questioned the AEC assertion that the blast could not trigger a large earthquake or tidal wave. Train concluded that the greatest danger was that the test might generate a large sea wave called tsunami.

The court of appeals Nov. 3 refused for a second time to halt the test. It said, however, that the case did "present a substantial question as to the legality of the proposed test."

Canadian protests—The scheduled Cannikin test had drawn strong protests from the Canadian government and public before the explosion

Among the major protests:

Public demonstrations were held throughout Canada Nov. 3-6. Thousands of Canadians demonstrated Nov. 3-4 at border crossings and blocked traffic at bridges leading to the U.S.

Eight Canadians Nov. 4 handed to a White House aide in Washington a 3,000-foot protest telegram allegedly containing 177,000 names of persons opposed to the test. Thousands of other telegrams were also sent to Washington.

The House of Commons Oct. 15 adopted (with only one dissenting vote) a resolution expressing "great concern" over the planned test and declaring its opposition to all nuclear testing.

Canadian External Affairs Secretary Mitchell Sharp Oct. 27 expressed to U.S. Secretary of State William P. Rogers a "sense of deep disquiet" over President Nixon's decision to proceed with the test [see above] and said Canada would hold the U.S. responsible for "any short-or-long-term" damaging effects of the blast.

Other opposition—More than 30 senators, headed by Edward W. Brooke (R, Mass.) submitted a petition to President Nixon Nov. 5 expressing opposition to the test. The telegram said: "We believe that to proceed with the test is to endanger national security and world peace, not to further it."

Sen. Mike Mansfield (D, Mont.) Nov. 4 denounced the scheduled blast as "an outrage" that "doesn't sit well with our allies, Canada and Japan." Sen. Mark O. Hatfield (R, Ore.) the same day submitted a telegram to Nixon signed by 5,000 constituents urging cancellation of the test.

A petition signed by 12 leading American scientists, including three Nobel Prize winners, Nov. 5 urged Nixon to delay the test to allow "full public and Congressional debate on the issue" before a final decision was made.

The head of the Environmental Protection Agency, William D. Ruckelshaus, said Oct. 2 that he was opposed to the test on environmental grounds.

Six leading scientists and arms control experts Sept. 24 sent a letter to Nixon urging cancellation of the test because of "physical and political risks that outweigh whatever possible advantages there might be to our weapons development program." Among the signers were Dr. George Kistiakowsky, former science adviser to the late President Dwight Eisenhower, James J. Wadsworth, former U.S. ambassador to the U.N. who helped negotiate the 1963 test ban treaty and Adrian Fisher, former deputy director of the Arms Control and Disarmament Agency.

Gas project radiation cited. The AEC Oct. 1, 1972 estimated that natural gas that might be recovered through underground atomic blasts would expose Los Angeles residents who would use the gas in unvented heaters to only 2 millirems of radiation a year and expose San Francisco users to 2.5 millirems. The accepted civilian exposure limit was 500 millirems a year, background radiation in Colorado constituted 200 millirems a year and one-time X-ray exposure constituted 50 millirems a year.

The AEC planned two more tests in its program to tap 300 trillion cubic feet of natural gas buried in Western rock deposits.

Colorado gas-freeing blasts opposed. The AEC and the Interior Department announced Feb. 9, 1973 that they would detonate three underground nuclear devices in Western Colorado to release deposits of natural gas.

The move was immediately criticized by representatives of three environmental groups, the Oceanic Society, Friends of the Earth and the National Intervenors, because of an alleged risk of radioactive waste dispersal.

The AEC said it would explode three devices of over 30 kilotons each more than a mile beneath the surface in oil shale deposits in Rio Blanco County, to create a giant well for gas collection. The agency said the project might eventually entail 280 explosions to create 60 wells to alleviate the shortage of "clean-burning natural gas." A "special explosive" had been designed, AEC said, to yield far less radioactive tritium than had been released in the earlier explosions in the series, called Gasbuggy and Rulison.

No date had been set for the explosions. The AEC was still studying consumer safety limits on tritium exposure.

Natural gas blasts detonated— Three 30-kiloton (equal to 90,000 tons of TNT) nuclear devices stacked in a steel well casing were detonated simultaneously May 17, 1973 more than a mile below the Piceance Creek basin in Rio Blanco County, in northwest Colorado.

AEC officials said initial checks after the blast indicated there had been no above-ground radiation leakage. The blast measured 5.3 on the Richter scale.

The experiment, called Project Rio Blanco, was jointly sponsored by the AEC and the CER Geonuclear Corp., a private Las Vegas concern. Cost of the test was estimated at $7.5 million. Future project plans called for opening the cavity created by the May 17 explosion in 3–6 months and "flaring" or burning off 300 million–800 million cubic feet of gas.

Environmentalists had attempted to halt the test with a suit charging it might contaminate water underground, fracture oil-bearing shale layers and produce gas that would be much more radioactive than normal gas. Colorado District Court Judge Henry E. Santo had rejected the arguments May 14 and authorized the AEC to proceed with the test.

Opponents of the blast were concerned that exploitation of the gas deposits in the Piceance Creek area would require an estimated total of 140–280 nuclear blasts, while 5,600–12,620 tests would be required to fully develop the gas fields under the combined area of Wyoming, Utah, Arizona, New Mexico and Colorado. Three hundred trillion cubic feet of gas were believed trapped in the five-state region.

EPA measures China A-blast. To underscore its concern about "the seriousness of atmospheric radiation," the Environmental Protection Agency estimated Dec. 27, 1976 that China's A-test explosion on Sept. 26 eventually could cause four cases of thyroid cancer in Americans.

There had been measurable contamination of milk in some Eastern states from fallout over cow pasture areas as the explosion's radioactive cloud drifted across the U.S.

EPA Administrator Russell E. Train emphasized that the health hazards "dramatize once again the need for an end to atmospheric testing of nuclear weapons."

Bikini Island contaminated by Radioactive fallout. The U.S. had decided to relocate 140 residents of Bikini Island to another island, the House Appropriations Committee subcommittee on the interior was told April 12, 1978. Radioactive fallout from a 1954 A-test had contaminated food grown on the island and made the island unsafe as a permanent residence, federal officials said.

The Interior Department in March had asked Congress to approve a $15 million relocation program.

The island had been the site of nuclear tests from 1947 through 1958. The Atomic Energy Commission in 1969 had declared the island safe for habitation, and some of the Bikinians, who had been moved off the island in 1946, returned.

Three years ago, however, medical tests started to show higher than normal amounts of radioactive strontium, cesium and plutonium among the residents.

The Interior Department currently imported food to the island, so that the Bikinians would not have to eat the contaminated domestic crops.

The timing of the resettlement depended, federal officials said, upon further medical tests to determine how acute the danger to the islanders was. Also, the government intended to conduct a radiological survey of nearby Enyu Island to find out whether it would be safe for the Bikinians. Enyu, part of the same atoll as Bikini Island, had been selected as the first choice for resettlement.

Interior Department official John De-Young said April 12 that Bikini Island would be "off limits for 30 to 50 years."

None of the current Bikini residents appeared to have become ill from the radiation on the island, Interior Department officials said. U.S. High Commissioner of the Trust Territory Adrian P. Winkel told the House Appropriations subcommittee that the Bikinians wanted to remain on the island "even knowing of the danger."

Winkel said that "there was some desire of other Bikinians [not resident on Bikini] to go there." Another official said that many of those who wished to go to Bikini wanted to take advantage of the free food program run by the Interior Department.

Even though the Bikini residents opposed resettlement, Winkel said, "it must be done for the absolute safety of them and their children."

A-Test Vet Wins Cancer Claim. The Veterans Administration's Board of Veterans Appeals ruled Aug. 1, 1978 that a veteran who developed Leukemia 20 years after he was exposed to radiation in the course of his military service had a service-related disability. The precedent-setting decision entitled Donald C. Coe to a higher pension than he was currently receiving.

Coe, a resident of Tompkinsville, Ky., had taken part in maneuvers conducted in 1957 at the Nevada site of a nuclear test a few hours after an explosion.

The appeals board decided that it was "reasonably probable" that the radiation Coe was exposed to then was a "competent causative factor" of the cancer that appeared 13 years after Coe retired from the Army.

The ruling was expected to lead other veterans to apply for increased benefits. About 2,400 former servicemen who had participated in similar maneuvers had reported recently to the Defense Department that they suffered from leukemia or some other form of cancer.

1945 A-bomb effects. To determine the genetic effects of the 1945 bombings, a report was issued Mar. 3, 1963 by Dr. Stuart C. Finch, Associate professor of the Yale School of Medicine. The report, based on the study of 70,000 pregnancies in Hiroshima and Nagaski found that (a) Among survivors, radiation increased the incidence of leukemia in 1951–52 to 30–50 times normal incidence; although the rate has declined since then, the total number of cases since Aug. 1946 was 2–3 times greater than average. (b) Radiation increased thyroid cancer incidence among survivors. (c) It caused some eye troubles in persons heavily exposed but caused no new diseases. (d) Radiation exposure of parents did not increase the incidence of congenital malformation among their children.

Index

DATE DUE

APR 2 1 '81	APR 5 '81		
AP 26 '82	APR 20 '82		
AP 26 '85	APR 26 '85		

DEMCO 38-297